MANAGING ENERGY RESOURCES IN TIMES OF DYNAMIC CHANGE

MANAGING ENERGY RESOURCES IN TIMES OF DYNAMIC CHANGE

WILLIAM H. MASHBURN

Published by
THE FAIRMONT PRESS, INC.
700 Indian Trail
Lilburn, GA 30247

Library of Congress Cataloging-in-Publication Data

658.2
M 39m

Mashburn, William H., 1928-
  Managing energy resources in times of dynamic change.

  Includes index.
  1. Energy conservation. I. Title.
  TJ163.3.M38   1988   658.2   86-46135
  ISBN 0-88173-035-1

Managing Energy Resources In Times Of Dynamic Change
©1989 by The Fairmont Press, Inc. All rights reserved. No part of this publication may be reproduced or transmitted in any form or by any means, electronic or mechanical, including photocopy, recording, or any information storage and retrieval system, without permission in writing from the publisher.

Published by The Fairmont Press, Inc.
700 Indian Trail
Lilburn, GA 30247

Printed in the United States of America

10 9 8 7 6 5 4 3 2 1

ISBN 0-88173-035-1   FP

ISBN 0-13-551326-X   PH

While every effort is made to provide dependable information, the publisher, authors, and editors cannot be held responsible for any errors or omissions.

Distributed by Prentice Hall
A division of Simon & Schuster
Englewood Cliffs, NJ 07632

Prentice-Hall International (UK) Limited, London
Prentice-Hall of Australia Pty. Limited, Sydney
Prentice-Hall Canada Inc., Toronto
Prentice-Hall Hispanoamericana, S.A., Mexico
Prentice-Hall of India Private Limited, New Delhi
Prentice-Hall of Japan, Inc., Tokyo
Simon & Schuster Asia Pte. Ltd., Singapore
Editora Prentice-Hall do Brasil, Ltda., Rio de Janeiro

TO JANE

CONTENTS

FOREWORD xi

1 ENERGY MANAGEMENT - A NEW DISCIPLINE ... 1
The Discipline Emerges
A Proper and Permanent Role
The New Opportunity
A systematic Approach
Planning and Motivating

2 ORGANIZATIONAL PLANNING 7
Objectives
Taking Stock
An Energy Policy
The Organizational Ladder
An energy Budget of Your Own
Reports, Reports, Reports

3 AUDIT PLANNING 23
Developing a Priority List
Pre-audit Planning
Conducting The Audit

4 EDUCATIONAL PLANNING 47
A Sea of Ignorance
A Plan for Educating Coordinators
A Plan for Educating Management
A Plan for Educating Employees
Hard work - Good Payback

5 EMPLOYEE INVOLVEMENT 59
The Untapped Potential
Basic Psychology of Motivation
Applied Psychology of Conservation
Ideas for Employee Involvement

6 CONTINGENCY PLANNING 75
Disruption - Earthquake or Squirrel
Seven Steps
Automate and Relax

vii

7   **STRATEGIC PLANNING** **83**
    The Final Step
    New Horizons For Energy Managers
    Strategic Planning Defined
    The Format

8   **WORKING WITH GROUPS** **91**
    Choosing - Group or Individual
    Characteristics of Group Members
    Workshops - A Valuable Tool
    Workshop Procedure - a Recommendation

9   **ECONOMIC EVALUATION** **97**
    Financing Schemes
    Life Cycle Costing
    Payback
    Internal Rate of Return

10 **FUTURE ENERGY COSTS AND AVAILABILITY** .. **127**
    Basic Facts
    Developing Your Own Crystal Ball

11 **PRESENTING YOUR IDEAS** **131**
    Who is Your Audience
    First Impression
    What Do You Want to Say
    Time And Tide Wait for No Speaker
    Wonder What This Switch Does
    Audio Visuals - A Helping Hand

12 **KEEPING THE PROGRAM ACTIVE** **147**
    Networking
    Workshop Activities

13 **TIPS FOR SUCCESS** **151**
    Five Tips to Success

14 UNDERSTANDING ELECTRICAL COSTS 155
Rate Schedule
Power and Energy - What's the Difference
Metering
From the Utility Standpoint
Reactive Demand
Analyzing Your Electric Rate structure
Analyzing Your Electrical Energy
Conducting An Electrical Survey

APPENDIX 193
A SAMPLE ENERGY POLICY
B THE ENERGY MANAGEMENT PROGRAM AT SOUTHWIRE
C PORTABLE ENERGY AUDIT INSTRUMENTS
D A SAMPLE AUDIT
E SOUTHWIRE'S STRATEGIC ENERGY PLAN
F INTEREST TABLES

FOREWORD

The purpose of this book is to provide a step by step procedure for implementing a comprehensive energy management program. Many books have been written on the subject of managing energy, but they have perhaps one introductory chapter on the subject of implementing a program, them jump right into technical topics. This one deals solely with the implementation.

It is written for that individual whose boss just walked by and said, " I want you to be our energy manager," and whose reaction was, "what is an energy manager."

It is written for the prima donna who has single handed saved thousands of dollars, but is now running out of ideas.

And, it is written for all in between.

Planning and motivating and managing aren't nearly as exciting to technically oriented people as is some of the new technology they encounter. But it is perhaps more important in trying to manage energy. So, I have tried to write this book in an easy style that will ease the pain.

It is not a book of theory. It is one that has been created from actual working experiences - not necessarily mine, but from successful programs. I am indebted to all who have had a role in helping make energy management a defined discipline. So is the rest of our country.

I am particularly indebited to Edward Stephan, now retired from DOE, who was most instrumental in getting me involved in energy management, and to Charles Dorgan of the University of Wisconsin who shared so freely with me in the early days.

Chapter 1

ENERGY MANAGEMENT - A NEW DISCIPLINE

THE DISCIPLINE EMERGES

The management of our energy resources did not begin with the historic oil embargo of 1973 as many people believe. Energy-intensive companies, particularly those in the chemical industry, were concerned with cost and availability long before the embargo. It simply brought the need to manage energy into sharp focus, and expanded it to all segments of society.

Managing energy shortly after the embargo may be described as a fire-fighting operation, during which those responsible were given much authority but often little direction. This era was followed by a conservation phase, motivated by the quick return on investment from accelerated high energy costs. Then, as things settled down and there was time for contingency and strategic planning, the discipline for managing energy began to evolve.

An examination of energy-related education programs can provide an insight into the development of this discipline, as well as the present status and future trends. For the past ten years, continuing education courses in energy management have been taught at Virginia Polytechnic Institute and State University. The enrollment for these courses has remained relatively high during this period, which is a strong indication that, at the professional level, energy management is still being incorporated into the overall planning even though the general public may at times become apathetic toward energy issues.

The audience makeup has changed during this period. The first classes consisted mostly of representatives from the industrial sector. Now there is more of a mix, with people from federal establishments, hospitals, utilities, universities, local governments, public schools, etc. The techniques that were developed for managing energy in industry are rapidly spreading to other segments.

Originally there was much activity at corporate level as programs were put in place. Once in place and operational, there was a decline in activity at this level. These changes were viewed by some as a major setback. I disagree. In reality it was a shifting of responsibility downward.

A PROPER AND PERMANENT ROLE

Energy management has now assumed its proper and permanent role, and must compete with other projects and programs on an equal basis. It is being incorporated systematically into business planning and into the organizational structure. At times, it may appear dormant, but if the structure is there, it can be fully activated to meet any emergency situation.

Energy is no longer in the spotlight, so managing it requires substantially more skills - especially in the areas of economic evaluation, planning, and motivation. In addition, new priorities are constantly emerging that get top billing and push energy further into the background. Most of the current priorities in industry are based on the need to meet foreign competition, so increased productivity and quality are major concerns at present.

Unlike most new concepts that emerge on the scene, bloom and fade away, the need to manage critical energy resources will remain. The economic return is there; price shocks will continue to be a part of the scene; the technology is changing so rapidly that it needs constant evaluation; planning for energy security will continue to be a challenge. All this adds up to a personal opportunity with a high degree of security for those considering hanging their hat on energy management as a career.

THE NEW OPPORTUNITY

There is now a new management culture making significant changes within industrial organizations. It was brought about primarily by the need to meet foreign competition in quality and production. This new culture is based on a set of human values that promotes not only the good of the organization, but the integrity and development of the individual. Much of the middle management has been eliminated in order to decrease the response time between top management decisions and creative ideas originating at worker level.

This makes the timing good to either initiate a new energy management program, or revise one that has become inactive. A good program depends heavily on the involvement of everyone in the organization, so it can supplement this new management culture, or can actually be a proving ground for the effectiveness of people involvement.

A SYSTEMATIC APPROACH

Energy management provides career opportunities in a unique way, in that it combines technology and management. There is much information available on energy technology, but very little on the management aspects. The purpose of this book is to provide a step by step procedures for organizing and managing an energy program, with emphasis on organization, planning, motivation, and establishing priorities. The procedures are applicable to all types of organizational structures - industry, government, university, public schools, hospitals, banks, etc,. Internal barriers and techniques for obtaining results will, of course, be different for each organization, but the basic principles are the same.

A recent opinion survey conducted by the Energy Managers Professional Council of the Association of Energy Engineers found that even though energy managers rated top management support high, availability of funds, lack of manpower and lack of energy awareness were cited as the biggest factors preventing their programs from reaching their potential.

The question then arises - whose fault is it that there is not enough funding, or manpower, and people don't understand the problem? We all tend to project the fault onto someone else, but the truth is, these are now responsibilities of the energy manager. The role of the energy manager has expanded as the discipline of managing energy has become more defined. The emphasis is now on the managerial aspect.

PLANNING AND MOTIVATION

Planning and motivation - these are the keys to reaching the full potential of an energy program. The discipline of managing energy has now matured to the point where a prima donna operating alone can no longer put an organization into a secure, competitive position. The program must be integrated throughout the whole organization with everyone contributing.

That's what this book is all about - planning, motivating and managing. These are not new terms - in fact, they may be worn and trite -but successful programs can and have been built on the techniques I will detail in the following chapters.

Why haven't I written it earlier? Because it had taken time through conducting the Energy Management Diploma Program, consulting experiences, and observing good working models to be able to formulate detailed plans that I feel confident will help you develop a good energy management program.

As energy goes through periods of dynamic change, the only way you will be able to properly manage it is with a well organized approach. Start now as you begin this book to think of yourself as an energy manager, not just an energy engineer.

Chapter 2

ORGANIZATIONAL PLANNING

Organizational planning for energy management should have two main objectives: 1) to establish a working structure for the program, and 2) to establish the authority to implement it. This planning should include the development of the management structure, an energy policy, an energy budget, and a reporting system.

In working with a number of organizations, both large and small, I find there is no standard organization for an energy management program. All have to be custom fitted into the existing structure and use available talent.

The Conference Board, Inc.- a global network of leaders who exchange information on management, economic and public policy issues - published Report No. 837 in 1983, giving the results of a research project in which they studied the energy management organizational structure of several large corporations. Their study came to the same conclusions; the energy management responsibility is determined principally by the company's existing organizational structure, and its management style. Therefore, the guidelines given here for organizing your own program are somewhat generic in nature. They are applicable to large or small organizations, and to any organized group attempting to improve on the way they manage energy, whether they be industrial, governmental, commercial, or institutional.

Many technical people have a tendency to work alone, whereas the success of an energy management program depends heavily on the involvement of others. This organizational planning process then is a good place to start with people involvement. One technique I recommend and use in working with groups is the Nominal Group Technique described in Chapter 8. I recently worked with a very large corporation to establish an energy management program. Using this technique with a select group of people, we were able to get valuable inputs on every phase of the planning process. It also helped pave the way for implementing the program by getting people involved, and making them feel they are a part of it.

Before beginning the organizational planning process, first take the time to find the starting point by determining what was done in the past. You can't determine where you are going if you don't know where you are.

TAKING STOCK

Very few organizations have done nothing about energy, so it is important to know the past history of these efforts. This is probably of greatest interest to the individual who has just been given the responsibility for energy management. Therefore, the situation analysis may consist of simply thinking through the past efforts, may be a formal report, or a combination of the two. The important thing is that it be considered in the planning process. Having an understanding of the attitude that now prevails throughout the organization as the result of past efforts is as important as knowing the technical accomplishments. Is the attitude throughout the organization positive, apathetic, or negative? Unfortunately, many programs became associated with discomfort as previous efforts to conserve energy concentrated on changing comfort levels.

One large corporation set up their program through the personnel department, and envisioned it as a public relations effort. As the energy manager became more knowledgeable in the discipline of energy management, he became more aggressive in his efforts. But a major barrier to the program was the way it was initially conceived.

If there is a good program in place and a new manager is taking over, it is important not to re-invent the wheel by doing, or attempting to do, things that were done before. There could be sensitivities associated with prior attempts that should be avoided, just as there may be successes that can be built on.

<u>Energy History at a Glance</u> A graphical display of an organization's energy usage for the past few years can portray their successes and failures in managing energy. It can also provide the base for establishing priorities, plotting progress, and managing the program. With today's computers and spread sheet programs, plotting this data is no longer the chore it used to be. In many cases, energy managers have found that the information they need is already stored in a computer.

Some of the things you may wish to graph are:

- Usage by energy types - Oil, gas, electricity, coal, propane - plot quantity and cost of all your fuels. You should also plot water usage since it is usually directly related to energy and is fast shaping up as the next critical resource to be managed.
- Convert all types of energy to Btu's - Graph individual types and total. Chart 2.1 gives typical Btu content for fuels.

Chart 2.1

Coal - 10,000 - 15,000 Btu/lb

Wood - 8,000 Btu/lb

Oil - 140,000 Btu/gal

Natural Gas - 900 - 1400 Btu/cu ft.

Gasoline - 130,000 Btu/gal

Kilowatt-hr - 3414 Btu/kw-hr

(Heating values of fuels vary widely. Consult your energy supplier to ascertain correct value.)

- Degree days - Keep in mind that degree days calculations are designed for residential structures and may not correlate with energy used in industrial or commercial buildings. However with some iteration, you may get an acceptable correlation. Figure 2.1 shows an attempt to correlate degree days with energy use for a small commercial building.
- Energy index - Establishing an energy index is more of an art than a science. Determining a single number to cover the multitude of variables is difficult, but can be valuable if the correlation can be achieved. This may be Btu's per pound of product if a manufacturing facility, or Btu's per square foot per year if a commercial or public facility.

During the present period when energy prices are stable, many energy managers simply track their progress by plotting energy cost. The problem with this is apparent when costs are fluctuating sharply upward causing total cost to rise in spite of decreased energy usage.

- Other indicators - You may wish to graph other things that

relate to energy use such as sales, production, number of people, etc., and see if you can get a correlation.

<u>What is the Potential</u> - From data accumulated and analyzed in the above steps, it should now be possible to determine where the emphasis should be placed on a program. Should there be a hard push toward energy conservation or energy security?

If little has been done previously, the push should be for conservation. The first 10% reduction in energy use can be readily achieved through low-cost/no-cost projects. On the other hand, some organizations have achieved as much as 50% reduction with a concentrated, well-organized effort. In this case, the concentration should be on providing energy security for the future through strategic planning as discussed in Chapter 7. Most organizations are somewhere in between the 10 and 50% reduction - but you need to know where yours is.

Selling the program - A technique for selling a program to management is to plot the organization's energy usage and cost for the past decade, then project these for the next five year. The instability of supply, and fluctuations in cost become readily apparent. While management is wondering how they will deal with future costs and instability, present them with a well-developed plan for managing energy.

Forecasting energy costs and availability is a task most of us try to avoid. The mortality rate for those attempting to make a career of this has been high. Still, it must be done, and the energy manager is probably the most knowledgeable person in the organization to make these projections.

Forecasts are generally a consensus of opinion, so the more averaged in, the better. Some analysts predict drastic structure changes in energy in the next two to four years, while others see the world as now better able to cope with spurious disruptions. The energy manager must be widely read, and have knowledge of the national and international energy situation in order to make good forecasts. Chapter 10 deals with making price and availability forecasts.

My recommendation for energy projections is that they be conservative, straight line, and for as short a time period as allowable.

AN ENERGY POLICY

SHOULD YOU HAVE ONE - A policy statement issued at top management level can be valuable in authorizing an energy management program - if the normal procedure of the organization is to use operating policies. Most do, but some operate with few or no formal statements from the chief executive. In this case, trying to buck the system to instigate an energy policy could be counter productive.

Many organizations that have had an energy policy for years, may find it to be out of date because of shifting emphasis on energy, or other related reasons.

Most energy management programs will benefit from a good, current policy authorized by the chief executive. It provides the authority for the program, and can allow the energy manager to play the role of good guy by helping to implement a policy imposed on the organization - even if he wrote it. It can help justify expenditures for such things as instrumentation, which is hard to do from an economic standpoint.

<u>WHAT SHOULD IT CONTAIN</u> - A policy should be brief - three pages at the most. Remember, this is not a procedures manual. If it becomes too detailed, then it also becomes restrictive.

Items that should be addressed in a policy are listed below:
- Objectives
  - Provide for the energy security of the organization for both immediate and long-range situations.
  - Utilize energy efficiently
  - Comply with government regulations - federal, state, and local.
  - Incorporate energy efficiency into new equipment and facilities.
  - Have a workable plan for accomplishing the above objectives.
- Accountability
  - Define top level responsibility and the position of energy manager.
  - Establish the position of coordinators.
  - Define and establish any committees or task groups associated with the organizational structure.

- Reporting
  - Establish the annual energy report.
- Training
  - Include a statement in the policy authorizing energy training at all levels.
- Method for Updating Policy
  - Because of the various circumstances that can outdate the policy, a method for updating should be incorporated. A sample policy is given in Appendix A.

THE ORGANIZATIONAL LADDER

Very few organizations have need for a full-time energy manager. Many start out full-time, but once the program is organized and in place, the time required to maintain it decreases. As a result, the energy management organization most often is a network superimposed on the existing structure. Therefore, there is no standard format for an energy management organization -it has to fit what is already there. So most energy managers wear two or more hats. This does not, however, decrease the importance of having a well-defined chain of command with assigned responsibility.

<u>WHERE SHOULD ENERGY MANAGEMENT BE</u> - I have found that energy management programs are located most frequently in one of the following departments: business planning, environmental, facilities, or engineering.

If there is a choice, I would rank them just as I have them listed. Business planning is the first choice because it is easier from that position to integrate energy strategic planning into the organization's overall strategic plan. Strategic planning is the last step. It can be accomplished only after all others are in place and operational. Chapter 7 discusses strategic planning in detail. Environmental responsibility combined with energy management is a very common structure. One subtle advantage is that dollars spent on environmental control have no readily computed payback, whereas, those for energy do. So, there may be a psychological lift for those involved because they now have an opportunity to contribute to savings.

Being located in the facilities operation generally provides many resources. The disadvantage is that there may be too many high priority projects that keep the energy manager from wearing the EM hat.

Engineering is probably the least desirable location because of the high priority of the work usually associated with that department.

Generally speaking, having the energy manager in a staff position is better than in a line position.

SELECTING AN ENERGY MANAGER The most important step in the formation of the organizational structure is the selection of the individual to be the energy manager. If you look behind any successful program, you will usually find one dynamic individual who doesn't wait for instruction, but has to be told when to stop. The rules may not be broken, but are often bent. This well-directed enthusiasm is important because top management may simply not have enough knowledge of energy management to give day-to-day direction.

In addition to enthusiasm the following qualifications are desirable for an energy manager:
- Varied technical background
- Experience in production process or plant engineering
- Business experience and management skills
- Communication skills - both oral and written
- Planning skills
- Training in the energy management discipline

Very few people will be found that have all these qualifications, so being trainable in those where they are weak should also be a consideration.

Most selections are made from persons in-house. The advantage is their familiarity with the technical aspects of the operation as well as the organizational structure and personalities involved.

COMMITTEE OR ADVISORY GROUP - Many energy management programs start out as a committee - or advisory group - appointed to do something about energy costs. Too often these groups spend most of their time trying to identify energy conservation opportunities (ECO's) when they should be developing a good energy management program.

The members are generally made up of representatives from various departments, so they may or may not have the technical skills to identify and prioritize ECO's, but can be a valuable asset in implementing a program within their department.

If the energy manager has the option of organizing and selecting the committee, I have two suggestions: First, establish a time limit for the period of service. Then you can rotate off the ineffective ones without offending them, and re-appoint those you wish to keep. Second, define their role as advisory, not decision making. Otherwise, the tail will wag the dog.

One effective way to utilize the committee is with the Nominal Group Technique - described in Chapter 8 - where the problem is defined, and a systematic approach is used to get potential solutions. A good place starting point is to ask the group, "What are the barriers to our energy management program?"

SHOULD YOU HAVE A TECHNICAL STAFF - Some corporate energy managers have available to them a technical staff that has the capability of conducting audits, doing economic analysis, and even the design.

If the overall program is dynamic enough that the lower echelons ask for this type of assistance, such a staff can prove to be valuable. Many times, however,- particularly in a new program,- the push is from the top down, and encounters opposition. The lower echelon may feel their autonomy is threatened, or that the staff is unfamiliar with their operation.

One way to overcome this type of opposition is to concentrate on those that will cooperate with you, then use an example of the work done for them to lure others.

COORDINATORS - THE KEY PEOPLE - Coordinators should be established throughout the organization down to department level. The selection of these individuals is very important. Many corporations have pushed the energy management responsibility down to company level, so they are now living with the selection made when the position was not so critical. A weakness in the overall program may result if these coordinators are not selected very carefully.

Managers who are not convinced of the worthiness of the program may assign their most expendable person, rather than their most capable. Some suggestions for preventing, or at least minimizing this, are:

- Establish minimum standards for qualifications
- Establish joint approval authority
- Convince the managers of the importance of energy management to their operation.
- Make this selection procedure a part of the energy policy.

A model for selecting and training company level energy managers was recently established by Vitro Tec Corporation of

Monterrey, Mexico. Their prime business is glass manufacturing, so they are an energy intensive industry. When they were recently informed by the government that the price for natural gas would double within the next eighteen months, a corporate energy manager was appointed, and with the training department, was given full authority to select and train company energy managers. Those selected were top engineers highly skilled in the various processes of glass manufacturing. A cooperative type training program was established with participants alternating between time spent in the classroom and time in the plant implementing the energy management program. Duration of the program was 14 months.

This amount of effort could probably not be justified by companies here in the U.S. unless our energy situation drastically changes, but it is a very good example to emulate. A well-trained cadre of people representing diverse skills will not only serve their own local organization well, but will be a technical resource from which to form audit teams, special energy task groups, etc., to serve the parent organization.

AN ENERGY BUDGET OF YOUR OWN

IDENTIFIED OR DEDICATED - Budgets for energy management programs fall into three categories: an identified budget, dedicated funds, and none.

Having an identified budget certainly adds to the prestige of the program. It also compliments the planning process, and facilitates accountability. Accountability is in the energy manager's favor because of the inherent payback potential of energy management. One company with a program that has been in operation with an identified budget for several years, recently calculated the return on investment for the life of the program. Having an identified budget allowed them to calculate this. Incidentially, this ROI was 500%

There are some disadvantages to having an identified budget. First, managing a budget is time consuming and takes away from the main task of managing energy. Secondly, an identified budget may place a top limit on the number of good projects that could be implemented. Whereas, a good entrepreneur operating without limitations might access unused portions of other budgets.

Most energy managers operate with dedicated funds that are part of some other budget.

<u>WHAT TO INCLUDE</u> - Items that should be considered in a budget for an energy management program are:

- Salaries
  - Full or part-time for the energy manager, technical staff, clerical help. Generally does not extend to others who may form a part of the overall program, such as departmental coordinators   . Training
  - Expenses associated with attending conferences and seminars, funds for in-house training
- Promotional
  - Publications, awards, etc.
- Identified Energy Projects
  - Projects that are to managed directly by the energy manager
- Energy Investment Funds
  - To be used as seed money or matching funds to other departments.
- Research and Development Projects
  - Usually limited to very large operations

ORGANIZATIONAL PLANNING 19

GETTING YOUR SHARE - There are some subtle tactics for showing the financial importance of energy that you may wish to employ for getting budget support. First, in order to determine the equitable number of people that should be included in an energy staff, set up the following ratio:

Energy Staff/Total Employees = Energy Cost/Total Budget

When you solve the ratio for the energy staff, the number usually comes out quite high - probably more than you need, and certainly more than you will get.

A second tactic is to compare energy savings to profit from sales. If the profit is 5%, then the number of dollars in sales required to realize one dollar profit is $1/.05 = $20. One dollar of savings then obtained through the energy program is worth $20 in sales.

REPORTS, REPORTS, REPORTS

Reporting is somewhat like forecasting. Most people prefer to avoid it. However, it is can make a contribution to the energy management program by providing the bottom line on its effectiveness. In addition, reporting generates visibility by periodically placing information about energy in front of key individuals. The report is probably of most value to the one who prepares it. It is a forcing function that requires all information to be pulled together in a coherent manner. The process of pulling it together and making it coherent requires much thought and analysis that otherwise might not take place. So, whether reporting is a requirement of the energy management program or not, my advice is to do it anyway.

WHAT KIND - The energy manager should, as a part of his overall planning, determine what kinds of reports will be needed to provide information to the decision makers who have an influence on the energy management program. Basically, there are three kinds of reports; project reports, monthly reports, and yearly reports.

Each project may have its own unique reporting requirements. It depends on the funding source, the total value, who is involved, and the type of project - just to mention a few. Government supported projects, for example, generally have the reporting format specified. An ice storage project supported by an electrical utility may have some special reporting requirements. In-house funded projects generally have fewer reporting requirements.

A project may or may not require interim reporting, but should have a final report. In addition, you should consider developing a one sheet project summary that gives a brief description, and payback. Such a summary can have many uses. Internally, they can be used to share information with other facilities. Externally, such a file can provide the basis of a technical paper, or can be a document of information that, in many ways, adds to the body of knowledge of energy management. Many companies now share this kind of information with others. It is good public relations. A monthly reporting of energy use can be the base from which all other reports - project, yearly, etc. - are generated. So, if it is well planned, all other reporting becomes much easier.

If monthly reports are supplied, a yearly report may or may not also be required. They are generally desirable where the activities of several facilities needs to be brought together in one summary report.

<u>EASING THE PAIN</u> - By making reporting a requirement of the energy policy, getting the necessary support can be easier. If energy reporting can simply be added onto other reporting requirements, there will also be less opposition. There is hardly an organization that does not already have some requirement for reporting. It may be production numbers, sales, number of students in attendance, etc. Having someone collect the necessary

energy data, and include that into the existing report then can be a simple task. So, one of the first steps in developing a reporting system for energy is to determine what kind of reporting is now being done for other activities within the organization.

Letting those who use the energy collect the data has a psychological advantage. They have a tendency to do something about it. So use this approach wherever possible.

The old saying that a picture is worth a thousand words is particularly true in energy. Have a 35 mm camera with you at all times, and record, preferably on slides, energy projects from beginning to end. In addition, photograph all energy saving techniques you encounter, as well as energy abuses. A large collection of such slides can have many uses.

KEEP IT SIMPLE - If the program is running smoothly, and energy is being saved, reports should be presented in a simple, concise manner that can be understood at a glance, because that may be all the time a busy executive has. If it is not going well, then give the report with a lot of numbers and statistics, combined with lengthy, technical explanations. Unfortunately, many good projects are presented this way.

The monthly report could be something as simple as adding to an ongoing graph that perhaps compares present usage to some baseline year. Try to keep the narrative length to one page. Have supporting data in your file so you can provide any additional in-depth information requested.

A slide presentation is one of the better ways to give an oral report, if the slides are selected very carefully to suit the audience. By having a large number of slides from which to choose, you can more easily organize the presentation to fit the audience. More information on giving a presentation is given in Chapter 11.

One company that has done an outstanding job at managing its energy is the Southwire Company of Carrollton, Georgia. John Kopfle, a Senior Energy Engineer with the company, presented a paper on their program at the 1987 Energy Management Diploma Program. I have included his paper in Appendix B.

Figure 2.1

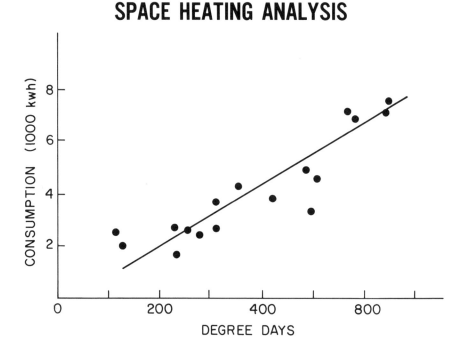

ENERGY AUDIT-COMMERCIAL
SPACE HEATING ANALYSIS

Chapter 3

AUDIT PLANNING

I have shown audit planning as the step after organizational planning. This is the logical sequence - to have the organization, the policy, the budget, and reporting system in place before trying to plan the auditing process. In reality, you may have to do some auditing to show the cost effectiveness of an energy management program in order to proceed. So, while this sequence is optimal, I recognize that it may be expedient to vary from it.

DEVELOPING A PRIORITY LIST

Energy auditing is an on-going thought process, supplemented by structured events, with the purpose of identifying ways to use energy more efficiently. The energy manager is constantly thinking of new ways to conserve or manage energy. To supplement this by a systematic approach involving others can clean a facility of energy waste.

There should be two objectives to auditing. The first should be to develop a priority list of energy projects. The importance of such a list cannot be over emphasized. In a field where "everyone is an expert," if you don't establish priorities, individuals or groups will do it for you - and they probably will not be the most productive.

The second objective should be to use the audit process as a training tool to further educate those involved in the program. It is a good opportunity to provide technical training as well as motivation. The energy manager of one large international corporation did this very effectively. He selected a set of portable instruments, packaged them to be transportable, then took them to each of the companies in the corporation. With the assigned energy coordinator, he conducted audits, and by using the instruments to take various measurements, taught them much about auditing. The technique proved to be very effective for his program.

It is much better to be invited in if you are auditing a facility other than your own. Getting an invitation may involve some of the motivational techniques you will find scattered throughout this book. Having a strong energy policy that is being enforced by top management is one of the better techniques. Once you are invited in, having local participation in the audit will increase the possibility of a successful effort.

If you are auditing an operation other than the one for which you have direct responsibility, your first question to management should be, "How will you implement the energy projects that will be identified as a result of the audit?" If there is no mechanism in place, then your first task should be to help get a program organized. Don't do all the work only to learn that the report is destined to collect dust on a shelf.

PRE-AUDIT PLANNING - DOING YOUR HOMEWORK

There is a tendency to go immediately into the facility and start looking for ways to save energy. However, if you will do some pre-audit planning, the actual audit will be much more productive. If the facility is very large and complex, such pre-audit efforts may take from three to six months to complete. I have nine recommendations for this pre-audit planning.

<u>WHAT HAS BEEN HAPPENING</u> - A review of the historical use of energy is the first step in pre-audit planning. If this was done in the initial situation analysis of the Organizational Planning, the data collected there may suffice for the pre-audit planning. As a minimum, however, you should have a record of the energy usage on a monthly basis for the past two years. This should be by fuel type, and include quantity as well as cost. A

AUDIT PLANNING 25

graphical display of this information makes the peaks, trends, and major users quite obvious.

A bar or a pie chart is useful in establishing broad priorities, such as major energy users - buildings, departments, process, etc. - major fuels used, and greatest costs. The pie charts shown in Figure 3.1 show the annual usage and cost of each type of fuel used in an industrial plant. From the chart, it is readily apparent that the greatest use is oil; however, by graphing costs, it is apparent that electricity costs more, and hence should receive first attention.

Figure 3.2 shows the percent of electricity used by various facilities in a camp resort. It is obvious that the recreation center should be the first priority for an electrical audit of the resort. Figure 3.2

Line graphs of electrical demand (kw) and electrical energy (kw-hr) can indicate areas of potential dollar savings. Figures 3.3 and 3.4 are typical graphs of these two functions in a manufacturing operation. Notice that they both have summer peaks, which indicates an air conditioning load. Priorities then, should be to look at improved efficiency of the air conditioning system, as well as the potential for load shedding.

As explained in Chapter 14, a very detailed analysis of electrical use can be made with just the data available on the electrical bills.

Your electric utility may be able to provide you with a much more detailed breakdown of your energy usage that can help in the detection of such things as peak loads of short duration. Figure 3.5 is a plot of natural gas usage that shows most going for heating.

It is time consuming to collect and analyze in detail the various uses of energy, but it can also be very productive. For example,

Figure 3.1
ENERGY SOURCES

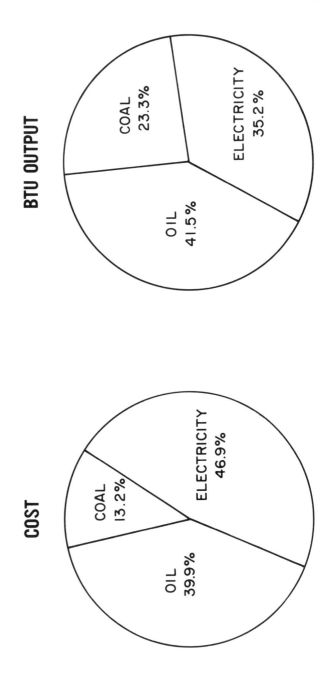

AUDIT PLANNING 27

Figure 3.2

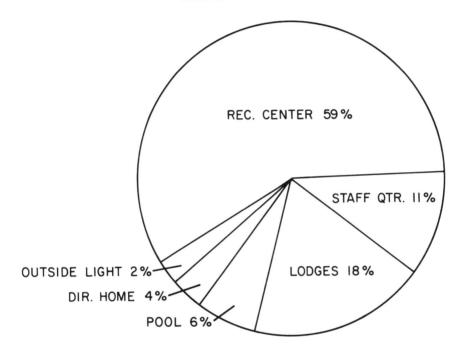

Figure 3.3

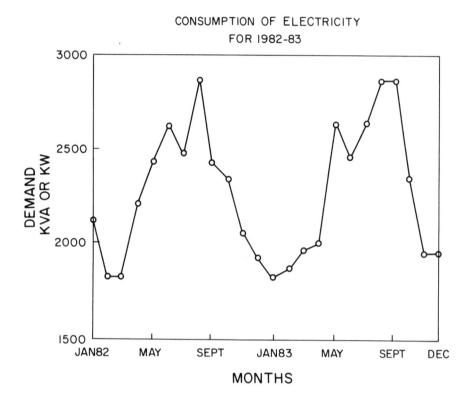

Figure 3.4

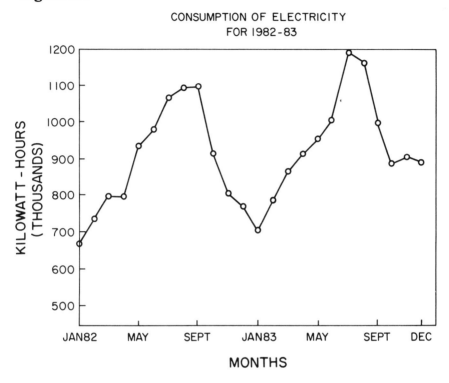

Figure 3.5

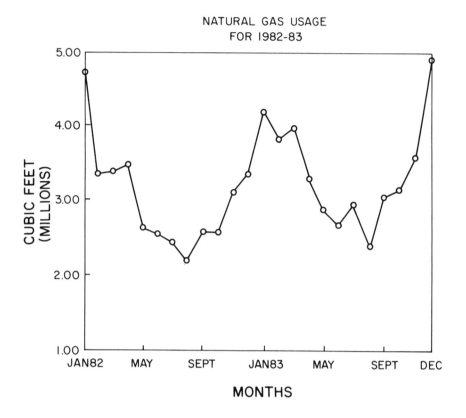

some states have a tax on energy, but will allow a deduction on that portion that goes into the product. This requires separating the energy used for environmental purposes - heating, cooling, lights, ventilation, etc. - from that used in the manufacture of the product.

The detailed breakdown may identify energy intensive products that should be evaluated for discontinuance, or at least proper pricing. A manufacturer of glass products recently, through the audit process, determined that one of their products used fifteen times more energy than any of the others. That particular product may be discontinued.

Water usage is another example where an analysis of its use may provide dollar savings. Many facilities pay a sewage bill based on the metered flow into the building, regardless of the internal use. For example, water that is evaporated in a cooling tower does not go into the sewage system, and hence should not have a sewage charge.

SINGLE LINE UTILITY DIAGRAMS - A very valuable tool for the audit team is a single line utility diagram, particularly if the facility being audited is large and complex, such as a manufacturing plant.

Figures 3.6, 3.7, and 3.8 are examples of line utility diagrams for electricity, steam and air in a manufacturing facility.

It takes much time and effort to prepare such diagrams, but they provide a graphical representation of the generation, distribution and consumption of each utility.

ENERGY RELATIONSHIPS - Another part of the pre-audit planning is a review of energy relationships - perhaps with the audit team. There are some square and cube root relationships that will allow substantially decreased energy usage if understood and applied properly. For example, lighting intensity varies with

Figure 3.6
ELECTRICAL POWER BALANCE, AVG. KW(%)

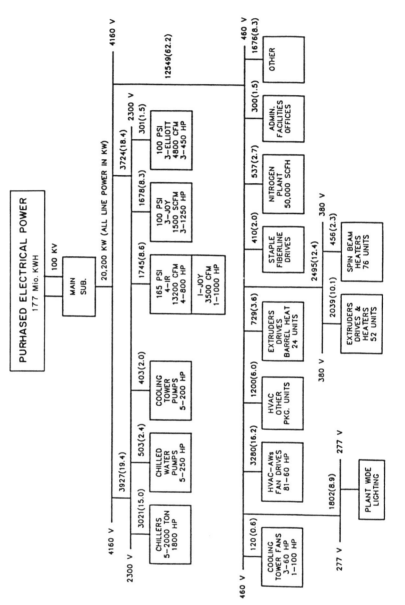

AUDIT PLANNING 33

Figure 3.7

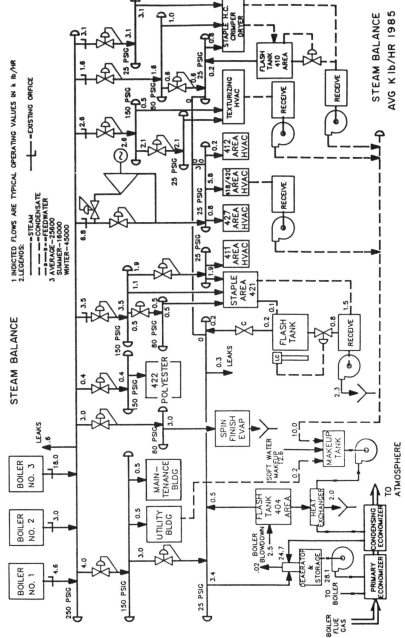

34 MANAGING ENERGY RESOURCES IN TIMES OF DYNAMIC CHANGE

Figure 3.8

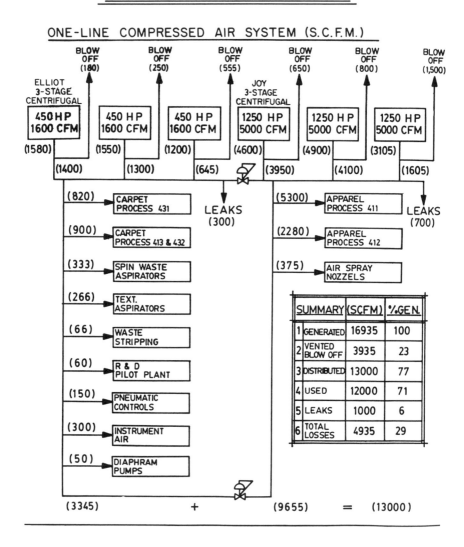

the square of the distance. This means that lowering the lights should have a high consideration in a lighting audit. Fan laws have a cube root relationship, so that reducing the CFM will reduce the power required by a cube ratio.

Chart 3.1 shows the energy relationships for some of the more commonly used applications. A familiarity with these will help in mentally establishing priorities when conducting an audit.

<u>WHAT IS THE COST PER MILLION BTU's</u> - A cost analysis of the various energy conservation opportunities (ECO's) requires correct information on the cost of energy of each piece of equipment involved. This makes it necessary to know the efficiency of the equipment. For example, if insulating a steam line will save 100 million Btu's over a given period, you must determine how much it costs to generate 100 million btu's. Therefore you must know the cost of fuel used, the Btu content of the fuel, and the boiler efficiency.

Chart 3.2 gives an equation for determining the cost per million btu's, and some typical seasonal efficiencies for various energy using pieces of equipment.

Determining the efficiency of process equipment may not be an easy task, but can be important, because many are very inefficient. For example, some drying processes have proven to have an efficiency as low as 4 to 5%. Before attempting to measure the efficiency of such equipment, check first with the manufacturer. They may have already made that determination. In addition, they may have some good ideas for improving it.

<u>AN ECONOMIC ANALYSIS PROCEDURE</u> - Having an established procedure for doing the economic analysis of the ECO's is necessary before starting the audit process. The reason for this is to make the economic evaluation of ECO's be compatible with that used by the parent organization. The procedure should be defined and put into a format that is easily understood, particular-

ly by the audit teams. Remember, you will be competing with other projects for available capital and other resources, so it is important that those who review the economic analyses see them in a format with which they are familiar.

Many organizations have a "hurdle rate" below which projects will not be considered for funding. Most often it is expressed as payback, but sometimes as return on investment (ROI). This hurdle rate will vary with the economic health of the organization, so you must have a method for keeping current. Generally, the more reliable the source of funding, the longer the required payback. Military establishments, for example, sometimes extend the payback to life of the equipment.

The first step, then, in developing the economic procedure should be to define this hurdle rate as it applies to energy projects. If there is no established hurdle rate in use throughout the organization, you should work with management to determine one for energy projects. On the other hand, if there is one in existence, it may or may not be the same for energy projects. In the earlier days of energy management, energy was so critical that it had a higher priority than other projects. In some organizations this high priority may still exist.

Before an energy project is finally submitted for approval, it will generally have been subjected to more than one type of economic analysis. Most start with a simple payback analysis, and if under a given dollar value, will require nothing more. Beyond that, they may require a more detailed study. These limits, then, need to be defined as a part of the economic analysis procedure. Finally, the type of economic analysis required by management should be determined. It may be internal rate of return - which is perhaps the most common - life cycle cost analysis, or some customized procedure. Someone in the Accounting Department can help you make this determination, and can usually make the analysis for you. Most have a computer program with the format and company accepted numbers programmed in. Just be sure to check those built in numbers on such items as energy cost, escalations, life of equipment, etc. They may not be updated, or correct for the project being considered.

Once the above information has been adequately defined, put it in a format that can be readily understood and used by the audit teams.

WHAT KIND OF AUDIT - Unless there is some external requirement for classifying audits - as some state and Federal programs have - I recommend you develop your own system. A good organized system will help those involved in auditing concentrate on the task, so a more thorough audit will be conducted. A shotgun approach may produce some good results initially, but cannot substitute for a planned procedure which will glean the facility for ideas.

Develop a list of the types of audits that would be effective in your facility, then give them a priority. You may wish to start the development of such a list by using the nominal group technique with one of your existing groups, such as an energy committee, advisory board, technical staff, or audit team. Some suggested types of audits are listed below:

- Tuning, Operation, and Maintenance (TOM) Audit - This is a good audit to start with because it can produce low-cost no-cost ECO's that will help to quickly establish credibility for the program. It should check to see that all operating equipment is at peak efficiency. The manner in which it is operated, as well as the time, is often overlooked in this audit, so emphasize the "operation" portion in the TOM. If there is not a preventive maintenance program in operation, consider initiating one just for the major pieces of energy using equipment.

- Replacement Audit - Focus on those things that can save energy by being replaced, either because it is not operating efficiently, or because of better designed equipment. This audit might include looking at such things as the following:

- Lamps
- Motors
- Steam Traps
- Valves
- Thermostats
- Roof Insulation

● Low Cost Installation Audit - Concentrate on those items that can have their efficiency improved by the simple addition of things like:
- Time Clocks
- Boiler Controls
- Equipment switches

● Controls Audit - Looking at the operation of existing controls on such things as HVAC, boilers, and processes could be one of your most productive, cost-effective audits. Very few controls operate as they were intended initially - even if they were properly designed in the first place.

You may need to have outside expertise to help with this audit, preferably someone not in the controls business.

● Utility Audit - Develop a separate audit for each type of fuel used in the facility - gas, oil, electricity, etc. Also, develop one for compressed air, water, and any other related utility.

● Engineering Audit - This type of audit is intended to be a detailed audit of the complete facility, and often involves the services of an outside consultant. The expertise of outside consultants is generally concentrated in HVAC, so if you are looking at manufacturing processes, your best help may be from in-house expertise.

● Employee Suggestions - A part of the employee involvement plan includes a procedure for them to submit suggestions for saving energy. (See Chapter 5) This is a good source for adding ECO's to the priority list.

● Micro-audit - One energy manager developed what he terms a "micro-audit". It consists, basically, of having a department supervisor and two employees spend twenty minutes at the end of each work week, looking over the department for ideas to save energy.

With help from the various groups in your energy management organization, you can develop other kinds of audits peculiar to your own facility that will allow you to generate a prioritized list of audits.

WHO IS ON THE TEAM - After developing the audit classification, the selection of audit teams becomes an easy task. Simply match the team to the type audit to be performed. For example, a TOM audit should include someone from plant engineering. For an electrical audit, you may wish to invite a representative from the electric utility. Most are now staffing with competent people, and are willing to assist.

If the coordinators in the program are competent and well trained, they make good team members. Another excellent source is energy managers from other organizations. In many cases, they may be more valuable than consulting engineers because of their broader experience.

In any case, make sure the teams have a good technical balance for the task to be performed.

COLLECTING THE ELUSIVE DATA - The kind of audit to be conducted will, of course, determine the amount and kind of data required. Obtaining this data is perhaps the toughest part of the audit effort. It can be difficult, expensive, and time consuming.

Once the kind of data desired has been determined, the sources that are available for obtaining it should be considered: Utilities - Be sure to check with your utility - gas, or electric - before trying to get data related to their services. They may already have it, or can get it easier, and less expensive than you.

They may also have an interest in monitoring a new piece of equipment you are installing. For example, many utilities now have an intense interest in thermal storage. They will not only instrument such a project for data collection, but may help in the initial funding.

- Manufacturers - Seasonal or operational efficiencies for specific pieces of energy-using equipment may be available from the manufacturer. This should be checked out before doing your own in-house testing of such equipment.
.Instruments - If you are taking data from instruments already on-line, be sure they have been recently calibrated. What often appears to be an efficient operation may in reality be a stuck meter. Stack temperature thermometers stuck for years at 350 degrees F. have been a good source of energy management stories. Recorded data that is consistent, or too good, may also be an indication of a malfunctioning meter.

While permanent instrumentation should be an overall objective of the energy management program, justifying and obtaining it are sometimes difficult. Renting or leasing should be considered viable alternatives for expensive instruments that will be used infrequently. Another option may be to purchase them at corporate level for use throughout the organization.

A set of portable instruments packaged for easy transportation can serve two purposes: It can be used for making required measurements, and can be incorporated into the training of the audit team. Appendix C shows a suggested list of instruments you may wish to incorporate into such a set.

<u>GET IT OUT OF THEIR SYSTEM</u> - If any members of the audit team are not familiar with your operation, consider giving them a tour of the facility prior to auditing just to allow them to view any new and interesting things that might distract them during the audit.

A CHECK LIST OF YOUR OWN - A check list specific to both the facility and the type audit to be performed can provide a lot of good ideas. Start such a list by first collecting all the check lists that are available to you from various sources. In the early days of energy management most of the publications on energy consisted almost exclusively of check lists, so there are many around. From these, select those applicable items, and start generating your own lists. Then continue to add to them from your own ideas, as well as from other sources.

Provide copies to your audit team prior to conducting the audit.

UTILIZE THE FREE SERVICES - There are many companies that will perform a free audit based around the product they are trying to market. For example insulation companies will audit your facility and identify the areas that will benefit from new or renewed insulation. So will manufacturers of steam traps, and energy management control systems, just to name a couple. You should develop a list of those free audits that are available to you and incorporate them in your overall auditing program.

CONDUCTING THE AUDIT

If the pre-audit planning has been thorough, and the team is well indoctrinated, the audit should produce meaningful results. There are three phases of the actual audit: (1) the identification of potential ECO's, (2) an analysis of those ECO's, and (3) the generation of a priority list of energy projects.

IDENTIFYING ECO's - This step involves the audit team's generation of potential ECO's by both analyzing the available data, and the walk-thru inspection of the facility.

Many people, in going through a facility, attempt to identify energy waste and develop solutions at the same time. Often the solutions are obvious, but for those that aren't, this technique may put a limitation on the number and quality of ideas generated. A better approach is to separate the identification and the solution process, so that identification takes place during the walk-thru, then solutions are developed by the team in a more relaxed environment.

If the audit team is large - say eight - you may wish to divide it into two groups of four. You may find it expedient to have someone in each group designated as the recording secretary. This prevents duplication of the effort of listing ideas, and also allows ideas to build on each other.

The people who work in the facility you plan to audit should be aware of your plans and supportive of them. You should exercise extreme care not to make anyone feel threatened by your findings. Many good ideas can come from this group is they feel positive toward your activities. This takes some prior salesmanship on the part of the energy manager.

ANALYZING ECO's After the identification of ECO's in the facility, the next step is to develop the best solution for each, along with the economic evaluation - which at this level is usually simple payback.

By having all team members involved in developing solutions, a larger number of quality ideas will be generated for each ECO, and from these, the best can be selected.

The amount of time and effort spent in this ECO analysis will depend upon the kind and complexity of the audit. It may require calculations, contacting vendors for more information, a second look at specific items, and other fact gathering before the process can be completed.

The end result, then, of this analysis should be a list of potential energy projects with a simple payback for each.

AUDIT PLANNING 43

THE END ITEM - A PRIORITY LIST - The priority list should be considered by the energy manager to be a "living list" that is constantly being added to from many sources - audits being just one. The objective of an audit should be to make a meaningful contribution to that list.

The priority list should contain brief but pertinent information on ECO's that have achieved the status of being considered for implementation. The following items should comprise the list:
Item No. - This is simply a numerical listing of the ECO's.

- Area - This can designate a broad location, such as "facility wide" for ECO's like steamtraps, or can refer to a specific location, such as "Boiler #3" where there is perhaps a leak.
- Unit/System - The piece of equipment involved in the ECO is listed here.
- Recommendation - The change to be made, or work to be done is described briefly.
- Estimated Cost - Total cost, including material and labor, is shown here. You may wish to show briefly what is included in the overall figure, such as "12 man wks plus mat'l."
- Estimated Yearly Savings - A brief explanation of this figure can be helpful, such as "2% of purchased energy bill." Also, if it is a one shot savings rather than a yearly, this should be noted.
- Payback - It is optional as to the time period you prefer, but many prefer months.
- Priority - If the list is long, you are probably better off to have four levels of priorities - such as A, B, C, and D - rather than attempt to prioritize the whole list. There are many factors that will influence the priority of a project. Cost and payback are usually the principle ones, but there are others that, at any given time, could be just as important, or more so - such as:
  - The need for high visibility with regard to energy. Many organizations installed solar systems, not because

of a good payback, but because it gave the impression that they were energy conscious.
  – The immediate availability of funds. This is particularly applicable with governmental agencies, where funding may suddenly become available for use given time frame, or there is a rush to spend unused funds at the end of the fiscal year.
  – Price changes, particularly with fuels.
- Responsibility - This may list an individual, or a department, but **it should be the one that has the most to gain.**

The priority list, then, is the culmination of the whole auditing process. Armed with it, the energy manager is shielded from spurious disruptions. Planning for implementation of energy projects can be integrated into overall planning by the organization. Projections of future needs can be made. The priority list can move the energy management program to a high, stable level. This is why it is so important to develop one.

Appendix D is an excerpt from an industrial audit showing some of the key components of an audit report, including a priority list.

CHART 3.1

ENERGY RELATIONSHIPS

Energy Content of Air
$Q = V * 1.08 * T_{diff}$
Where Q = heat loss in Btu/hr
V = rate of exhaust in CFM
T_{diff} = temperature difference

Energy Content of Water
$Q = 500 * V * T_{diff}$
Where Q = heat content in Btu/hr
V = flow rate of water in GPM
T_{diff} = temperature difference

Conductive Losses
$Q = U * A * T_{diff}$
Where Q = heat losses in Btu/hr
$U = 1/R$ = coefficient of transmission in Btu/hr/sq-$_o$F
A = area in sq ft
T_{diff} = temperature difference

Fan Laws
$HP2/HP1 = (RPM2/RPM1)^3 = (CFM2/CFM1)^3$
Where HP = horsepower
RPM = rev per minute
CFM = cubic feet per minute

Pump Formulae
Pump shaft horsepower
$HP = HQ*(SG)/3960*np$
Where H = pump head in feet
Q = flow in GPM
SG = specific gravity in lbs per cu ft
np = pump efficiency
Electrical demand
$KW = HQ/(5300* np * nd)$
Where KW = kilowatts

H = pump head in feet
Q = flow in GPM
np = pump efficiency
nd = drive efficiency

Light Laws

$E = I/L^2$

Where E = illumination in foot-candles
I = intensity in candle power
L = distance in feet

Electrical Power

$KW = E * I * \cos O$ (single phase)

Where E = voltage
I = current
$\cos O$ = power factor

$KW = 1.73 * E * I * \cos O$ (three phase)

Chart 3.2

PROCEDURE FOR DETERMINING ENERGY COSTS PER MILLION BTU'S

Equation for determining cost per million Btu's:

$$\frac{\$}{MMBtu} = \frac{\$}{\text{unit of energy}} \times \frac{1}{\text{seasonal efficiency}} \times \frac{1}{\text{MMBtu/unit of energy}}$$

ENERGY CONTENT OF MOST USED SOURCES

Source	Typical Seasonal Efficiency	Approximate Energy Content In
Oil - No. 2	65%	.140 MMBtu/gal
Oil - No. 4	65%	.149 MMBtu/gal
Oil - No. 6 (heavy sulfur)	65%	.152 MMBtu/gal
Oil - No. 6 (low sulfur)	65%	.144 MMBtu/gal
Gas	70%	.10 MMBtu/therm
Electric	95%	.0034 MMBtu/kwh
Heat Pump	175%	.0034 MMBtu/kwh
Coal	65%	25 MMBtu/ton

EXAMPLE:

Cost of #2 fuel oil = $1.10/gal
$/MMBtu = $1.10/gal X 1/.65 X 1/.140 MMBtu/gal
        = $12.09/million BTU

Chapter 4

EDUCATIONAL PLANNING

A SEA OF IGNORANCE

Energy has been a subject of major interest in this country for over a decade, **but there is still a sea of ignorance concerning the status of the various energy sources, and the technology involved in managing them.** Because both the technology and the energy situation are changing so rapidly, only those with an intense interest plus reliable sources of information are in a position to keep informed. Therefore, a major part of an energy management program has to be the educational effort.

Raising the energy educational level throughout the organization can have big dividends. The program will be a smoother operation if management understands the complexities of energy; the committee will work on a much higher, more productive level; coordinators will be kept abreast of the latest technology; the quality and quantity of employee suggestions will take a quantum leap.

Everyone in the organization, should be considered a candidate for energy education, and it should be a "continuing education," not a one shot effort. Generally there are three audience classifications that should be involved: management, coordinators, and employees. Although there is a lot of overlap, a separate educational plan should be developed for each group. Below are some suggestions for educational plans.

A PLAN FOR EDUCATING COORDINATORS

This group should receive the most attention in educational planning because they are the most involved.

A two or three day seminar held annually can be the base for the whole educational effort, and the general theme should be planned for at least the first three years.

The first seminar can be a make or break, so plan it very carefully. Get some help from others, either energy managers or educators, who have been involved in successful seminars. Use in-house resources. If there is a training department, get them heavily involved. There are a lot of details involved in such a program, so use all the experienced help available.

STRAIGHT FROM THE TOP - **This is your best opportunity to get, and exhibit, commitment from top management - which is a necessity for any program.** So, invite the highest level executive in the organization to give a few opening remarks. But, find out before hand what he or she intends to say. If the chief executive officer walks out and commits the program to something you hadn't planned on, you may be in trouble from the beginning. In order to prevent this, you may offer to write the speech, or at least provide some material for inclusion.

A SHOT IN THE ARM FOR COORDINATORS - Making the coordinators believe that their involvement in the program has a personal payback is important at this stage. Many of them may be there because they were told they were to be the coordinator for their department, and not because they volunteered. Their workload may make them resentful of the time away from their present job, so a major objective of this first seminar is to convince them that they will experience new learning, and will develop skills in a discipline that is opening up new opportunities.

Many will come there feeling that their task of being an energy coordinator is overwhelming, and that they have unique problems. Workshop activities that deal with these problems can do much to dispel these insecure feelings. In addition, they often are a source of good ideas to use in further developing the overall program. So build two or more workshops into this first seminar.

In every program, try to give the attenders something to take back and implement immediately. This can usually be done by including technical training on some aspect of energy that is common to most - if not all - of the attenders. One or two good ideas can sometimes pay for their time in the seminar. This makes them, as well as their supervisors, feel good about the whole effort.

<u>IT DOESN'T COST MUCH MORE TO GO FIRST CLASS</u>- **Make the seminar a first class event.** It reflects on you and the energy program. Have good professional speakers, whose appearance and presentation will enhance the program. Top the seminar off with a banquet. If you decide on an after dinner speaker, make it light entertainment, not another technical subject. Too many after dinner speakers at such events do not realize that the participants have already had a full day of listening to technical topics, and they ramble on - and on.

A manual should be prepared in advance and given out at the beginning. It should include a schedule of events, a bio-sketch of the speakers, a list of attenders, information on each topic presented, and other things that will help pull the whole seminar together. Maps of the city, a restaurant list, and a listing of places of interest are always good for out of town participants. If it is an international group, such things as the dollar exchange rate will be helpful.

You may wish to have a logo developed and introduced at the seminar. Again, any available professional help, or talent should be utilized to design the logo. Small favors with the logo on them - such a cups, carrying cases, etc.- can help introduce the logo.

Some makers of energy products may be willing to donate door prizes that can be given out periodically during the seminar.

THE FIRST YEAR - This first seminar should have as its main theme the structure and operation of the energy management program. The role of each person in the program should be made clear, as well as the services available for assistance to them. Procedures should be explained, and future plans discussed.

In addition, you may with to include basic information on more technical subjects, such as economic evaluation, understanding electrical costs, etc.

Chart 4.1 contains suggested topics and workshop activities from which you may wish to select in preparing your course content. These are taken from outlines of successful seminars that other organizations have had.

One technique that I use in a seminar is to have a "round robin" session, where each person is given an opportunity to either tell about their overall program, describe a successful project, or solicit advice from the group on an energy problem or project. This will take a lot of time. With fifteen people, you should allow one half day. with a larger number, you may with to break them into smaller groups of like interest. It is a very good way to expose the in-house resources, and get the group working together. Past evaluations of seminars by participants show this activity to rate very high.

THE SECOND YEAR- For the second annual seminar, you may with to concentrate on sharing program results. This is best achieved by having selected participants give a presentation on one of their successful projects. This approach has several positive actions. First, the person giving the presentation gains experience in presenting a project report, and receives recognition from the peer group. Second, by concentrating on the successes, others are motivated to become more active. In addition, the internal resources for managing energy are being further developed and made known throughout the organization.

It is not a good idea to require every participant to give a presentation. Many may not be able to give one of the quality you want for your seminar. Some companies have tried this, but realized it was a mistake.

In addition to the individual presentations, you should add some technical presentations, perhaps by vendors that have done a good job for your organization, or are recommended by some other reliable source, such as another energy manager.

Some motivational topics should always be included in the seminar. In their workplace, the coordinators may be surrounded by a sea of apathy, so providing a shot of inspiration is important. In addition, having a credible presentation on the future cost and availability of different fuels can keep the broad perspective in front of the coordinators.

THE THIRD YEAR- The third annual seminar will occur after the program has reached some level of stability, so there is more freedom to place the emphasis on the most suitable topic. It may be a mixture of topics including technical subjects, motivation, and energy projections. By this time, the program should be well along toward contingency and strategic planning, so energy projections become more important.

A PLAN FOR EDUCATING MANAGEMENT

The yearly seminars can be a vital part of your effort to educate management people in energy matters. You may wish to invite a select group to the first half day of the seminar, then design the program so that those topics of most interest to them are presented during that time.

It may be difficult to arrange a special meeting with management just for the purpose of discussing energy, but you may be able to give a presentation on an appropriate energy topic as a part of another meeting. So, find out about these regular meetings and plan to become a part of them. If there is a special topic that you feel should not share the limelight with other subjects, then develop a special presentation for management. You may wish to include others in the program if they can make a contribution. Remember, you are "managing" the energy program, so don't try to show expertise in every subject yourself. Your ability to identify and use experts may be a better reflection on your capabilities.

Chapter 11 provides suggestions on giving a presentation. It is important that you not degrade the whole energy program by a poor presentation.

The energy report provides a mechanism to inject educational tidbits by including some additional explanation, where applicable, in the report. Very short articles that are pertinent to your educational goals, taken from magazines and newspapers can be attached to reports and sent selectively.

A PLAN FOR EDUCATING EMPLOYEES

The potential contribution to an energy management program by employees is so important that I have included a separate chapter on their motivation and involvement. Educational planning is dealt with to some extent in that chapter, but I will emphasize some points here because educational planning and employee involvement should really have separate plans - even though there are some commonalties.

A two or three day seminar, similar to that held for coordinators - but with a different emphasis - is a good way to start employee education. This is a heavy commitment of resources, but has proven to have a good payback. If there should happen to be a cutback in working hours, or a temporary shutdown, try to get that time allotted for such training.

During such a seminar, you may wish to teach the fundamentals of the energy systems in which the employees are involved, such as electrical, steam, air, etc. However, one topic that you should surely include is energy conservation in the home. Studies have shown that if employees are taught how to save energy in their homes, this training and attitude will carry over to the workplace. Your state energy office may be able to provide you with handout materials relating to energy conservation in the home.

The value that employees place on educational efforts in their behalf cannot be overemphasized. I recently participated in a three day seminar for employees at a large corporation. In a workshop setting, we asked them to prioritize the things that motivated them to save energy. Every group listed an educational experience as either their number one priority or very near the top.

Chart 4.2 lists some topics that you may wish to include in a seminar for employees.

In addition to the seminar, there are other techniques that should be considered in an overall educational plan for employees. Many manufacturers now have good presentations - obviously biased toward their products - that can be used to educate employees as to their use and application. Energy topics can often be injected into other regular meetings, such as safety. These topics should be developed somewhat like a soap opera, in that they have continuity. This will keep interest high.

If there is an in-house organ, there are a couple of options you may use as part of your educational plan. The first is to prepare - or have prepared - a series of articles on pertinent energy topics for inclusion in the paper. The second is to develop a correspondence course for employees. This might consist of a series of articles in the paper, a test, and a certificate for those satisfactorily completing the course.

HARD WORK - GOOD PAYBACK

This whole educational effort is a lot of work. This is a good time to re-emphasize the fact that you should "manage" the energy program, not attempt to do it all. If there is a training department in you organization, make this their prime responsibility under your direction. If not, solicit talent from other sources to help you. But you must do the planning.

Chart 4.1

ENERGY SEMINAR TOPICS FOR COORDINATORS

1. The Energy Management Discipline
   - From Firefighter in the 1970's to Strategis Planner in the 1980's
   - Reasons for Permanency of Energy Management
   - Skills needed
   - Obtaining Skills
   - Where Energy Managers Fail
2. Projected Energy Costs and Availability
   - Domestic Projections
   - Global Projections
   - Impact on Developing Nations
   - Use Trends and New Technology
3. Steps in Developing an Energy Management Program
   - Establishing a Data Base
   - Employee Involvement Plan
   - Organizational Structure
   - Consultants, Manufacturers and Utilities
   - Audit Plan
   - Economic Evaluation Policy
   - Energy Information resources
   - Priority ECO List
   - Educational Goals
   - Reporting System
   - Contingency Plan
   - Strategic Plan
4. Energy Audit Plan
   - Types of Audits
   - Energy Relationships
   - Measurement and Instrumentation
   - Team Selection
   - Simulated Walk-thru Audit with Slides
5. Economic Evaluation Techniques
   - Financing Methods
   - Basics of Life Cycle Costing
   - Hurdle Rates

6. Understanding and Reducing Electrical Costs
   - Demand, Energy, Reactive Power
   - Rate Structure
   - Analysis Techniques
   - Motors, Lighting
7. Employee Involvement Plan
   - Basic Psychology of Motivation
   - Motivation for Energy Conservation
   - Specific Ideas for Employee Involvement
8. Contingency Planning
   - Utility Plan
   - Energy User Identification
   - Curtailment Priority
   - Shutdown Plan
   - Alternative Supplies
9. Strategic Planning
   - The Process and Purpose
   - Information Sources
10. Incorporating Preventive Maintenance into the Energy
    - Management Plan
    - Basics of Preventive Maintenance
    - Selective use for Energy Management
    - Developing a Partial Preventive Maintenance Program
11. Energy Information Resources
    - Associations
    - Publications
    - Energy Management Network
12. Keeping an Eneragy Management Program Active
    - Networking

WORKSHOP TOPICS

- What are the Barriers to the Energy Management Program
- What is the Role of the Energy Coordinator
- Where Should the Energy Program be Five Years From Now
- What Techniques Can Be Used to Motivate Employees to Save Energy
- Develop a Priority List of Energy Conserving Ideas

Chart 4.2

ENERGY SEMINAR TOPICS FOR EMPLOYEES

1. Projected Energy Costs and Availability
   - Domestic Projections
   - Global Projections
   - Impact on Developing Nations
   - New Technology
2. Energy Conservation in the Home
   - How Heat Flows
   - Understanding R Values
   - Conservation Priorities and Techniques
   - Heating Systems
   - Hot Water
   - Electrical Energy Conservation
3. Understanding Electrical Energy
   - Demand and Energy
   - Power Factor
   - Motor Operation and Controls
   - Lighting
   - Heating
4. Steam Systems
   - Why Use Steam
   - Boilers
   - Steam Traps
   - Insulation
5. HVAC
   - Types
   - Controls
   - Conservation Techniques
6. Industrial and Commercial Buildings
   - Building Operation
   - Window Treatment
   - Air Leaks
   - Ventilation Requirements
7. Energy Audits
   - Simulated Walk-thru with Slides

SUGGESTED WORKSHOP TOPICS

- What are the Barriers to Conserving Energy
- What is your Role in Managing Energy
- Develop a Priority List of Energy Conserving Ideas

Chapter 5

EMPLOYEE INVOLVEMENT

Part of the job of an energy manager is selecting a strategy to suit the particular times or circumstance. At this writing, energy prices are relatively low. Consequently, less money is being budgeted for energy conservation projects. Now is the time to concentrate on low-cost no-cost energy projects. And the best source for these is the employees.

In my opinion, involving employees in the energy management program is one of the greatest untapped sources for making a major contribution. These are the people who have an intimate knowledge of the operation and environment of energy systems - but seldom have a mechanism for contributing to the program.

In my experience in working with employees, I have found that most suggestions they submit are ones they have known about for a long time. Nobody had taken the time to ask, or to listen - or, more importantly, provided them with the motivation to think about and submit ideas for conserving energy.

Employee involvement can have a good payback, but before you jump into it, you should have a basic understanding of the psychology of motivation, particularly with regard to energy. If you ask about employee involvement, too many times you will be referred to a poster somewhere in the building that says "Save Energy", and that will be the extend of it.

Before entering seriously into a program involving employees, be prepared to give a heavy commitment of time and resources. Plan it. Don't just give it lip service. Energy managers with a well organized program will devote an average of 20% of their time to working with employees.

This chapter is divided into three major sections. The first deals with the basic psychology of motivation - what it is, who has it, how people establish priorities.

The second section is a summary of a study by two psychologists from the University of Colorado - Cook and Berrenberg [1] - for the Department of Defense, who was trying to determine the effectiveness of some of their motivational programs. It deals specifically with motivation to save energy.

The third section discusses specific ideas that you may wish to select from in your own program. They are taken from several programs in which I have been involved, and which seemed to have merit.

BASIC PSYCHOLOGY OF MOTIVATION

Motivation may be defined as **the amount of physical and mental energy that a worker is willing to invest in his or her job.** Three key factors of motivation are listed below:

- Motivation is already within people. The task of the supervisor is not to provide motivation, but to know how to release it.
- The amount of energy and enthusiasm people are willing to invest in their work varies with the individual. Not all are over-achievers, but not all are lazy either.
- The amount of personal satisfaction to be derived determines the amount of energy an employee will invest in the job.

It has been my observation with people, that once you get to know them, you find there is one outstanding thing they do, either at work, as a hobby, or in a second job. A machinist may play a musical instrument in a blue grass band. An assembly line worker may invest in real estate and could buy and sell most of us. A worker in the Physical Plant Department may be the state champion muskie fisherman. Truth is, I know these three people.

They may not write a book on energy, but I don't play a musical instrument, never make money through good investments, and would love to know how to catch muskie.

Because people work at menial tasks, don't think it is because they are lacking in brain power. They may simply not want the responsibility that goes with executive type jobs. There is certainly something to be said for being able to leave your work at a work table when you go home at night.

Achieving personal satisfaction has been the subject of much research by industrial psychologists, and they have emerged with some revealing facts. For example, we have learned that most actions taken by people are done to satisfy a physical need - such as the need for food - or an emotional need - such as the need for acceptance, recognition, or achievement.

The research has also shown that many efforts to motivate employees deal almost exclusively with trying to satisfy physical needs, such as raises, bonuses, or fringe benefits. In this day when most people have a fair wage and other benefits, these methods are no longer very effective - except for the short term.

We must then look beyond these to other needs that may be sources of releasing motivation.

Psychologist Abraham Maslow [2] has suggested that needs fall into five categories that form a ladder of priorities. He also suggests that these needs must be met in sequence. That is, if a need lower on the ladder is not satisfied, a person will be motivated to satisfy that need first and will be "blocked" from releasing higher levels of motivation until it is met.

Another important thing that Maslow says is that once a human need has been satisfied, it no longer motivates a person. Another then takes its place. Figure 5.1 shows Maslow's rankings.

SELF-ACTUALIZATION:
Full development of abilities; creativity fulfilled personal life

ESTEEM NEEDS:
Self-respect and the respect of others

SOCIAL NEEDS:
Sense of belonging; groups membership; love; acceptance by others

SECURITY NEEDS:
Absence of threats to life, health, and safety; orderly environment

PHYSIOLOGICAL NEEDS:
Food, shelter, and clothing; environment that sustains life

FIGURE 5.1 THE NEEDS LEVELS OF HUMAN MOTIVATION

A number of studies by different organizations have verified Maslow's priority listing. In one such study by Hersey and Blanchard [3], both workers and supervisors were given a list of ten job-related factors. Supervisors were asked to rank them in the order in which they thought these were important to the workers. Workers were asked to rank them in order of importance to themselves. The table of Figure 5.2 shows how each answered the questions.

Job Related Factors	Supervisors	Workers
Good working conditions	4	9
Feeling "in" on things	10	2
Tactful discipline of workers	7	10
Full appreciation for work done	8	1
Management loyalty to workers	6	8
Good wages	1	5
Promotion and growth in the company	3	7
Understanding of personal problems	9	3
Job security	2	4
Interesting work	5	6

FIGURE 5.2 RANKING OF JOB-RELATED FACTORS

Note that the factors the workers ranked 1-2-3 were ranked 8-9-10 by the supervisors. we often overlook the fact that most workers have enough of life's necessities and enough job security to be reasonable well satisfied with these job-related factors.

Job enrichment then is the key to motivation.

APPLIED PSYCHOLOGY OF CONSERVATION

This section is primarily a review of a paper written by Stuart W. Cook and Joy L. Berrenberg entitled "Approaches to Encouraging Conservation Behavior; A Review and Conceptual Framework." [1] It is published in the 1981 issue of Journal of Social Issues.

In the paper, they discuss seven categories of approaches to encouraging energy conservation. The following is a review of these.

Persuasive Communications

Underlying this approach to encouraging conservation behavior is the premise that such behavior will occur more often among those believing in the need for conservation and in the results of conservation practices. Although studies show conflicting results with respect to the existence and strength of such an association, it is felt that beliefs in the need for conservation are important. This seems rather obvious, but as we explore this further, we find there must be more than a belief in conservation before any action is taken.

Considerable attention has been given to the question of how much fear should be aroused by the persuasive communication. There appears to be a positive relationship between fear arousal and persuasion: the higher the level of fear induced, the greater the change in attitudes and behaviors. However, it seems to be most effective if the fear appeals deal with topics primarily of significance to the individual, e.g., personal well being.

Attempts to instill the fear of a national energy crisis as a method of promoting attitudes favorable to the conservation of energy may involve disbelief and rejection for a number of reasons. First, evidence of excess profits by oil companies implies that a crisis is manufactured rather than actual.

Second, individuals are confident about the ability of science and technology to develop sources of energy other than fossil fuels.

The success of persuasive communication is directly related to the credibility of the source of communication and may be reduced if recommended changes deviate too far from existing beliefs and practices.

Conservation Behavior

Studies have shown that even though persons have a pro-conservation attitude, they are no more likely to be taking conservation actions than those without such an attitude. The reasons for this are many and complex, but there are some approaches to encouraging conservation behavior that are discussed below.

Directing Attention to Conservation: Conservation programs direct attention to available opportunities for conserving resources in several ways. The most prevalent is via localized signs or short messages in the mass media. When such reminders are very general (e.g. "Conserve Energy") they have little effect. Two features of reminders appear to increase their effectiveness: (1) display the reminder at the point of action and the appropriate time for action; and (2) specification of who is responsible for taking the action and when it should occur.

Public Commitment: Studies have shown that pro-conservation attitudes and actions will be enhanced through associations with others. Such associations may include the presence of others with similar attitudes, the pro-conservation statements of such people, including group leaders, and events such as making public commitments to conservation.

Material Incentives and Disincentives

Incentives and disincentives that may be employed to encourage conservation include financial regards and costs, increases and decreases in comfort and convenience, social approval and disapproval, commitments to cooperate and share with others, etc.

The central assumption underlying the use of incentives to foster conservation actions is that rewarded behavior is repeated. However, the effect of any given reward upon performance depends in part upon the existence of corresponding need in the person. (Refer to earlier section on Maslow's ladder of priorities).

As a resource shortage grows, a society resorts increasingly to disincentives rather that incentives to control consumption. However, the impact of disincentives on behavior is complex and unpredictable.

Financial Incentives: In our society, we have come to depend heavily on financial incentives to motivate desired behavior. Studies show that positive effects are achieved with financial incentives. Some of the things that tend to decrease the effectiveness of financial incentives are:

- An incentive that is small in relation to the cost of the resource.
- If the socio-economic stature of the consumer is such that the monetary incentives represents a small increment in spendable income.

Convenience/Comfort Incentives: Consumers place considerable importance on the potential inconveniences of discomfort of reducing their consumption of energy. This indicates the desirability of increasing, when possible, the convenience and comfort associated with conservation.

The use of disincentives becomes more frequent as a resource crisis deepens. People react against compulsory changes that involves them without their consent, so disincentives should be used only under conditions in which consumers may be certain that curtailment of resources is necessary and is being equitably shared.

Social Incentives and Disincentives

Behavior is motivated by social as well as material incentives, but the use of social incentives in encouraging conservation has been infrequent. Three uses of social incentives are described below.

Providing Social Recognition and Approval: Social recognition occurs through such things as the award of medals, designation of employee of the month, and selection to membership in elite sub-groups.

Publicity for such social rewards is given through the posting of photographs, letters from the company president, articles in newsletters, etc.

The widespread use of such recognition testifies to the general confidence in its motivational efficacy.

Seeking Public Commitment: When we commit ourselves to important people to carry out a valued activity we have put ourselves in a relationship in which approval/disapproval from them may be anticipated. There is evidence that this type of commitment increases the likelihood that conservation actions will follow.

Involvement in Group Conservation Decisions: The potentially most powerful source of social incentives for conservation behavior - but the least used to date - is the commitment to others that occurs in the course of group decisions.

Models of Conservation Behavior

Research has shown that the endorsement by prominent people of products, political candidates, organizational programs, health practices, etc. does have a positive influence on behavior modification in our society.

Models are needed who are likely to be emulated since satisfaction is derived from promoting a feeling of similarity to the model.

Implementing Conservation Intentions

As a result of one or more of the preceding approaches, consumers may be motivated to conserve energy. However, there are still a number of reasons why they may not do so.

Lack of knowledge: The carrying out of energy-saving actions may be blocked by the lack of knowledge necessary to recognize and implement appropriate actions. Without proper training, people simply do not know what the priorities should be for reducing energy. In the home, for example, many will turn out lights as their major energy reducing effort.

But when compared with other energy using devices, lighting ranks very, very low.

In an industrial or commercial environment, this lack of knowledge of proper action is even greater. This is why I place heavy emphasis on employee training as a part of the energy management program.

Availability of Conservation Alternatives: A person may not have the opportunity or access to conservation action. For example, a person may not car pool because he or she does not believe it would be easy to find a car pool match. They might not put the most energy efficient heating system in their home because of lack of funds.

Anticipated Negative Consequences of Conservation: A third factor that may inhibit intended actions is the anticipation of negative consequences. For example, individuals may be hesitant to reduce air conditioning because they are concerned with the negative consequences of heat on their personal health and comfort. Conservation campaigns in such cases should concentrate upon showing how satisfactory cooling might be obtained without air conditioning.

Information on the Effectiveness of Conservation Efforts

Feedback Procedures: Once conservation actions have been initiated, emphasis then shifts to whether or not such actions will persist. It is generally agreed that the consequences of performance govern its continuation or termination.

Most consumers have little or no knowledge of the level and rate of their energy consumption. Many individual short-period conservation actions result in savings too small to be noticeable. Certain types of feedback make such savings evident.

Feedback to be effective must attract attention and be interpretable by the consumer.

Self-monitoring Feedback: A disadvantage of obtaining feedback is the cost involved. This can be eliminated by having consumers themselves collect the data.

IDEAS FOR EMPLOYEE MOTIVATION

Two very important things must be kept in mind when considering the implementation of any employee involvement plan. First, there must be a commitment to time and resources to carry it to completion. For example, if an award system for energy conservation suggestions is implemented, be prepared for the increased work load necessary to make quick responses. This generally requires additional staffing on a temporary basis. Summer student help can be of value.

Second, recognize that most motivation schemes have a limited attention span. Estimate this span and have a planned termination date. Don't let it dribble out and die.

Based upon the foregoing studies by psychologists as to what works and what doesn't, and upon personal experiences, the following ideas are submitted for your use where applicable.

Educational Plans: Cook and Berrenberg reaffirm that a lack of knowledge is a major barrier to implementing energy conservation ideas. Many people simply do not know what to do or the priorities for doing them. For example, people will make an all-out effort to reduce the energy used for lighting in their residence, but on a priority listing of energy uses, lighting is so small as to be listed at "other".

Education is something that is highly valued by employees. This may not be fully recognized by supervisors.

In a recent workshop on energy for employees at a large industrial plant, we decided to find out from the group what things motivated them to conserve energy. Using the nominal group technique - described in Chapter 8 - we asked them to provide a prioritized listing of things that provided motivation to conserve energy.

Of the seven groups, six of them had as their top two suggestions, things that related directly to education and group activities involving idea exchange. Chapter 4 has further information on energy education for employees.

Energy Committee: As stated in the section under social incentives, the potentially most powerful source of social incentives for conservation behavior - but the least used - is the commitment to others that occurs in the course of group decisions.

After employees are educated on energy matters, committees can be very effective for both motivation and the development of ideas.

EMPLOYEE INVOLVEMENT 71

Before they are formed, however, careful planning should be done to assure effectiveness and maintain interest.

Again, the nominal group technique is recommended because it allows everyone to participate on an equal basis. Problems dealt with should be specific. Rotating the chair position is a good method of developing leadership and interest.

Awards Program: Positive effects can be achieved with financial incentives. This at first appears to be contradictory to Maslow's priority of needs where physiological needs are at the bottom. The difference is probably attributable to the fact that one-shot financial rewards don't get built into the family budget, but are used for self rewards.

If the award is small in proportion to the value of energy saved, the effort may not be effective.

One technique that has been used successfully is to combine a financial award with a lottery, so there is a chance for a substantial award. This also creates the opportunity for social recognition through the publicity of the winners.

Simulators: Simulators have proven to be effective as teaching tools. One company now produces a unit that can be programmed to simulate the energy use in a residence. By varying individual components - such as inches of attic insulation - the cost difference is immediately displayed.

The energy manager of one company purchased an energy simulator and then developed a motivation program for employees which he entitled "Give Yourself a $50 per Month Raise."

With the use of the simulator, employees first determined their present energy costs. Some training was given to employees along with literature on energy conservation available from the state energy office. Six months later, they again determined their energy costs. The average savings achieved at that point was $35 per month.

This program has several positive features: energy education, financial reward, and social incentives.

There is a strong indication that when people are shown how to save energy at home, the attitude for saving energy will carry over to their work place.

Micro-Audits: Micro-audits are short (20 minutes) audits usually conducted at the end of the week by line supervisors and some of the employees. Their chief value lies in maintaining the energy conservation effort. For example, they may determine if a light switch installed for a specific purpose is still being used.

Such audits can be used as a motivation technique by involving employees on a rotating basis.

CONCLUSION

The use of employees in an energy management program can be very effective if a few basic psychological facts are understood and incorporated into the planning. Proper involvement of employees requires a heavy commitment to time and resources, but can have high payback because most suggestions generated by them are in the low-cost no-cost category.

REFERENCE LIST

1. *Cook, Stuart W.*, and *Joy L. Berrenberg*, "Approaches to Encouraging Conservation Behavior: A Review and Conceptual Framework," Journal of Social Issues, Vol. 37, No. 2, 1981, p. 73-107.

2. *Maslow, Abraham*, Motivation and Personality, Harper and Row, 1970.

3. *Hersey, Paul*, and *Kenneth H. Blanchard*, Management of Organizational Behavior: Utilizing Human Resources, 3rd ed., Prentice-Hall, Inc., 1977, p. 47.

Chapter 6

CONTINGENCY PLANNING

The planning discussed in the previous chapters is tactical planning intended to be used in the day to day operation of managing energy. If this planning has been done properly and implemented, you are now in a position to begin looking at the future by thinking about energy security.

There are two considerations that must be given to providing energy security. The first deals with the "what if" situation. What if this energy supply becomes unavailable; what if this boiler fails. This is the type of disruption that occurs unexpectedly, or at least within a short time frame. The contingency plan is designed to minimize the impact of such disruptions. Developing such a plan is the subject of this chapter.

The second consideration of energy security is the strategic plan, which looks at energy on a broader scale and on a longer range. This is covered in the following chapter.

DISRUPTION - EARTHQUAKE OR SQUIRREL

Energy disruptions are almost always regional in nature. In this country, for example, the gasoline shortages of a few years back were scattered, with the large metropolitan area experiencing the worst. The natural gas shortage in 1979 caused school closings in some parts of the country, while others were virtually untouched.

Disruptions can be caused by a great variety of reasons, either man-made or natural. The man-made disruptions includes such things as explosions, prolonged strikes, equipment failure, etc. We might even include those made by animals. There have been many disruptions of electrical service caused by squirrels and birds.

Contamination by toxic materials is becoming an increasing source of disruption, not only of energy sources, but of the whole operation. Transformers containing PCB have caught fire on several occasions causing not only electrical outage, but also prolonged evacuation of facilities.

Natural disasters are a frequent source of supply failure - hurricanes, floods, and earthquakes being the most common. In severe winters, ice on the inland waterway may block coal barges for a sustained period of time, to the extent that many utilities relying primarily on coal began to phase in parts of their contingency plan.

Contingency planning on the international scene opens up an additional set of problems that we in this country do not as yet have to face. In some of the developing countries, electrical power is so limited that its availability may vary on a daily basis. One day, it may be available for a whole shift. The next, it may be cut off in the middle of the afternoon - with very little prior warning.

The potential for sabotage and acts of terrorism have to be considered in some parts of the world. Companies there, where terrorism is a real threat, may have the electrical power supplied by two separate lines as a part of their contingency planning.

WHY PLAN

The best planning in the world cannot anticipate all the things that can happen to cause energy disruptions, but it can minimize the detrimental effect.

One writer on the subject of contingency planning stated that "Energy is to production as breath is to life." Very few organizations can continue to operate very long without a sustained source of energy, whether they be industrial, commercial, or governmental.

Because of the multitude of sources for disruptions, the probability of one occurring is relatively high. So time is well spent in contingency planning.

Most disruptions occur without prior warning, so the reaction time is short. Without a plan such disruptions can sometimes have a "snowballing" effect that can shut down the whole operation.

Contingency and strategic planning for energy is done on a national level for both security and economic reasons. If this planning transcends to the lowest level echelon within the country it will add even more to national and economic security.

EIGHT STEPS

There are eight basic steps to developing a contingency plan. These are listed and discussed below.

1. <u>Know the Utility's Curtailment Plan</u> - Each utility serving your organization - whether electric or gas - has its own curtailment plan. You are somewhere on that hit list, and your first priority is to find out where. Most utilities don't like to admit that they have some vulnerability, but, if pressed, will drag out their curtailment plan and share it with you - particularly when they learn that you are developing one, because yours may benefit them also.

The curtailment plan is different with each utility. It is dependant to a great extend on their mode of operation, fuel sources, reserve capacity, etc. This means that the contingency plan you develop for one facility may not be the same as that needed for another that is served by a different utility.

Chart 6.1 lists the steps in the contingency plan for two different utilities, one in the Southeast and the other in the Midwest. An asterisk indicates those actions that will affect the users. Notice that the southeast utility has a series of internal actions before user becomes involved, whereas the Midwest utility jumps right on the user.

2. <u>Identify Energy Using Equipment</u>- Each item of equipment that uses a form of energy, such as electric, fuel, natural gas, or propane should be identified and listed. One potential source for such a list is the preventive maintenance program - if there is one.

Your list should contain the energy using piece of equipment and the type and amount of energy it uses.

Electrical energy is usually consumed by lighting, motors and air conditioning systems. The various lighting loads are readily determined by counting the number of lights and multiplying by the power consumed by each light - which is written right on fluorescent lights. One horsepower in an electric motor is equal to approximately 1KW of electrical power. However, most motors are oversized by an average of 30%, so you may wish to derate this further by multiplying by 0.7.

The rate of consumption of fuel oil or natural gas - or both - used for heating should be determined on a per degree day basis, so that you can project your requirements on a daily basis. This is vital in a curtailment situation when looking at reserve capacity.

3. <u>Develop a Shutdown Priority</u>- The shutdown priority should be developed in close coordination with the contingency plan of the serving utilities, particularly with respect to step function reductions. For example, the Midwest utility has a 30% reduction by industrial users as the fourth level of their curtailment plan. If this were your utility, you should have this step reduction built into your contingency plan, so when the word comes to implement this reduction, there is no major disruption in then trying to decide how to achieve it. However, for your master plan, you may wish to have a planned reduction in smaller increments, perhaps 10 %, that would meet other contingency needs. It should

be continued in these increments until production feasibility is lost. Following are some suggested ways to achieve cutbacks:

- Reduce lighting loads
- Turn off hot water heaters
- Turn air conditioning systems to fan only.
- Turn thermostat settings to minimum levels.
- Re-schedule certain operations to off-peak periods.

4. <u>Analyze Alternate Energy Sources</u>- Developing sufficient back-up energy sources, in most cases, will be very difficult. However, the alternative may not be acceptable either. To begin the analysis, each facility or operation should be reviewed with the ultimate objective of determining the cost effectiveness of having an alternative source of energy to sustain that particular operation. One large textile operation installed cogeneration equipment based solely on the expense of downtime if their electrical utility had a power failure. The revenue from the generated electricity was a side benefit. This is not uncommon with continuous production operations.

Every organization has a minimum point below which it cannot go without sustaining losses inappropriate to the cost of some type of alternative energy equipment. It may simply be to protect buildings from freeze damage.

Some of the things that may have already been installed for economic reasons may now be considered as back up systems for the contingency plan. For example, boilers that have dual firing capacity between gas and oil were probably done so to allow fuel switching in order to leverage prices of the two. Economic justification may have been the reason for the installation of cogeneration equipment that may now be incorporated into the contingency planning.

Reserve storage capacity for oil or propane may have contingency planning as its prime purpose, but may pay for itself through purchases when prices are lowest. If you are planning to add such storage, the price advantages of being able to purchase fuel by the complete tanker size should be considered.

So, even though providing alternative sources for supplying energy seems prohibitively expensive at first, using it in a dual role may make it economically feasible.

5. <u>Develop an Employee Lay-off plan</u> - Employee lay-off during the period of curtailment should coincide with the operation that is affected, whether it be production equipment or office space. This plan should then be cleared by the personnel department, any unions involved, and the people affected. The last thing you need in an emergency situation is to have a union on your back. Also, the people that will be affected should, out of consideration for their personal inconvenience, be informed of the plan, and should be alerted as far ahead as possible of any impending shutdown.

6. <u>Determine Complete Shut-Down Level</u>- As the song about the gambler says, "you need to know when to hold them and when to fold them." As the shut-down procedure progresses, there will be a cross-over point when it is more economical to close the whole operation than to try to sustain it. This cross-over should be determined in the contingency plan.

7.<u>Evaluate Your Suppliers</u>- During the natural gas shortage of 1979, I received a call from a manufacturer asking if I knew of a company that could heat treat a small component for them. Their supplier had no natural gas, and the component was critical to their product. They had provided for their own energy security, but had failed to check the security of their critical suppliers. Make this a part of your contingency plan.

8. <u>Annual Update</u>-Because you are dealing with a very dynamic plan, a part of it should be an annual review and update. The addition of new facilities, equipment, products, people, -all can begin the obsolescence of the contingency plan. If not updated on a systematic basis, it soon will be totally obsolete, or at least require a major effort to restore it. Treat the plan as if it were a part of the preventive maintenance program. Don't wait for it to fail before you repair it.

AUTOMATE AND RELAX

After all details of the contingency plan have been worked out and approved, you may then wish to automate it as much as possible to remove some of the human element. If you have an energy management computer system, much can probably be programmed into it, particularly the shutdown priority described in step 3.

Chart 6.1

SOUTHEAST UTILITY CURTAILMENT PLAN

* 1. Interruptible loads
  2. Supplemental oil firing
  3. Auxiliary boiler firing
  4. Emergency hydro
  5. Extra load capability of units
  6. Curtailment of generating plant use
  7. Curtailment of non-essential building load
* 8. Voltage reduction
* 9. Curtailment of special interruptible loads
  10. Curtailment - sales to other utilities
* 11. Voluntary load curtailment
* 12. Mandatory load curtailment

MIDWEST UTILITY CURTAILMENT PLAN

* 1. Voluntary cutbacks
* 2. 5% reduction of line voltage
* 3. Additional 5% reduction
* 4. Industrial users - 30% curtailment
* 5. Progressive reductions
* 6. Rotating interruptions

*   Action affects users

Chapter 7

STRATEGIC PLANNING

THE FINAL STEP

Strategic planning is the final step in the planning process of developing an energy management program. In reality, I have found very few organizations that have progressed this far with their program. It seems that once some of the major energy problems were solved, the thrust stopped there. This short-term thinking may cause serious problems if we are again thrown into turmoil with energy supply and costs. Many of those that have not planned ahead may not make it - or as a minimum will re-enter the management by crisis phase.

Some type of planning for the future is surely already taking place somewhere within your organization. It may be called strategic planning, long range planning, a five year plan, or some similar designation. Find out what is now being done, and start the process to have energy included.

The main purpose of this future planning is universal with all organizations; it is to provide energy security and maximum efficiency in the use of energy.

NEW HORIZONS FOR ENERGY MANAGERS

Not only does this planning expand the program, but also the role of the energy manager. Through this planning process, the energy manager is automatically projected into a range of broader responsibilities. For example, the following responsibilities may emerge:

- Analyzing facility closing and relocation options
- Reviewing the design of new facilities and/or operations
- Leasing and buying captive energy sources

- Analyzing the product energy cost of competitors
- Identifying less energy intensive substitutes
- Developing energy efficient manufacturing substitutes

STRATEGIC PLANNING DEFINED

Richard H. Cooper, Jr., in his career with R. J. Reynolds Industries, Inc., has served as the corporate Energy Manager. In that role, he developed an energy strategic plan for the corporation that has served as a model that many others are now emulating. Most of the information in this chapter is taken from the material he uses in his presentation on strategic planning at the Energy Management Diploma Program held annually at Virginia Polytechnic Institute and State University.

While there are many forms of future planning, Mr. Cooper recommends Energy Strategic Planning as the best method for incorporating energy into the business plans of the organization.

In trying to define strategic planning, let's look at what it is -- and is not. Cooper says "strategic planning is a thought process -- necessarily reduced to paper for communications purposes. The purpose is to provide guidance for current decision-making through the identification of key issues affecting the future of an organization and the establishment of effective strategies for addressing those issues."

Below are Peter Drucker's observations on what strategic planning is not:

- Strategic planning is not forecasting. In fact, strategic planning is necessary because we are unable to forecast beyond a short time span with any degree of precision.
- Strategic planning does not deal with decisions that are made in the future. It deals with decisions that are made today that will affect the future.
- Strategic planning does not eliminate risk. It helps management weigh the risks that it must take.

THE FORMAT

Cooper suggests a format for a plan that I have modified slightly by moving "Energy requirements and Costs" from third to second place, and listed some items separately. It is as follows:

> Executive Summary
> Energy Requirements and Costs
> Situation Analysis
> Issues
> Thrust or Objectives
> Strategies
> Action Programs
> Proposed results
> Sensitivities and Assumptions

Each of these will be discussed in the following sections, with examples.

<u>Executive Summary</u>- This is the last part of the plan to be completed, but will be the first section of the written plan. It should contain the essence of the strategic plan in a summarized and concise manner.

Major points that should be included in the summary are; an in-depth discussion of the first years plan, highlights of major commitments, and highlights of the major sections of the plan.

<u>Energy Requirements and Costs</u>- If an analysis of energy usage and future projections was made in the initial organizational planning stage, much of this data can be used here. Many energy managers find that all the necessary data has already been col-

lected and stored somewhere in a computer, and is simply a matter of retrieving it and putting it into a useful form.

My suggestion is to put it into the following graph format:

Pie charts -   . % consumption by end use

. % cost by end use

. % consumption by energy source

. % cost by energy source

Line graphs - a line graph of each of the above, projected for the next five years - or whatever the time projection is for your strategic plan.

In projecting energy requirements, planned or anticipated changes in facilities or production levels should be considered.

If you have a reliable energy index, such a Btu's per lb. of production, you may take the projections from the production department to arrive at your energy projection. Simply multiply their quantity by your energy index, then add overhead energy.

Figures 1 thru 4 in appendix C show examples of the above charts and graphs.

Situation Analysis - or Where Will We Be in Five Years-This review is somewhat similar to that recommended in the organizations planning section, except the emphasis is on looking forward to see where the organization is going, rather than looking backward to see where it has been. It also involves looking at external factors as well as internal. Simply developing a list of these factors, however, adds little to the strategic planning process unless their impact is assessed.

Some examples of external factors are:

- Supply outlook for various fuels - The supply outlook for most fuels looks good for the next few years, with natural gas perhaps being the most turbulent market. Is there a natural gas "bubble" that will eventually have an adverse effect on your operation? Does your electric utility have a planned phase in of additional generating capacity that will cause step increases in your electrical costs? What will be the impact of such changes?

- Government - Regulations and Policies - At this writing, acid rain legislation is impending. Presidential elections will occur within the five year planning period. What impact would these have on your operation?

- Technological Changes - Will your manufacturing process remain competitive through the planning period? If not, will you update equipment or product line? Is your present energy management control system already obsolete?

- Shifting Markets - Will the number of people using your present facilities increase, decrease, remain the same? Will your share of the market grow, decline?

- Internal factors can be either a strength or a weakness. Some factors that you may wish to consider for you own operation are:
    - Management commitment to the energy management program
    - Strength of energy management program
    - Alternative fuel capabilities
    - Supplier agreements
    - Energy storage facilities
    - Status of technology being used
    - Energy intensity of product

Issues- Key issues are identified by analyzing the external factors and the internal strengths and weaknesses of the organization. They should be matters that can seriously affect the organization if not resolved.

One issue that is common to all organizations is the threat to the continued availability of traditional forms of energy, and the possibility of supply interruption. On a more regional basis is the forecast of significant step increases in the cost of electrical energy.

Objectives- After the key issues are defined, then objectives must be established that will allow strategies to be developed. The objectives should be stated and be in measurable terms to facilitate later evaluation.

Strategies- Strategies may be defined as broad actions designed to accomplish objectives. Each objective may have one or more strategies.

Action Programs- Action programs are the working end of the whole strategic planning process. This is where the planning is done to take specific actions to implement each strategy. There may be one or more action programs for each strategy. Each one should include the work steps and their scheduled completion, and the personnel and financial resources required.

Proposed Results- This section should discuss all the benefits to be derived from the Strategic Plan, and should be quantified to the extent possible, so that reporting will be made easier, as well as evaluation.

Sensitivities and Assumptions- Recognizing the various factors and events that can influence your planning, as well as being able to clearly define the assumptions upon which the strategies are based is perhaps the most important step in the Strategic Planning process. The Proposed Results then, may take the form of a range of numbers rather that one. It may be a graph with a band rather than a single line.

The accuracy with which these sensitivities and assumptions are built into the plan may determine whether or not the plan needs to be updated yearly. Tracking and measurement are difficult when plans change from year to year. By the time one year's actual results are in, the plan has been revised. Opportunity lies in focusing on objectives and strategies that tend to remain stable even in periods of rapid change.

Each sensitivity should be discussed in a paragraph which relates back to a planned objective and strategy. They may apply to more than one.

MODELS TO GO BY

Figure 7.1 is a graphic display of the strategic planning process.

Southwire Company of Carrollton, Georgia has developed a very good strategic plan for their company using the format developed by Cooper. Appendix E is a copy of their strategic plan.

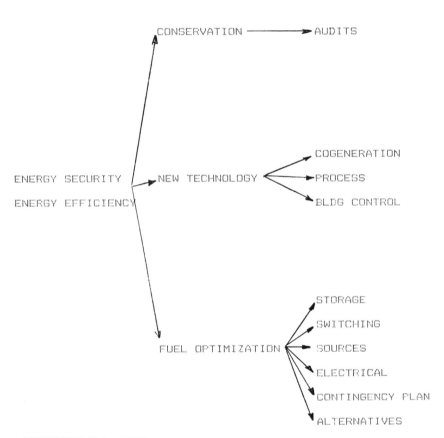

FIGURE 7.1 STRATEGIC PLANNING PROCESS

Chapter 8

WORKING WITH GROUPS

Moving from a prima dona stage to working with groups may a major step for some energy managers who have traditionally worked alone or with a small technical group, and have concentrated on the technical problems. But, as mentioned previously, if the program is to reach its full potential, it must expand to include all segments of the organization, and requires the ability to utilize group dynamics in the process.

There have been many belittling comments about committees, such as "The camel is a horse designed by a committee." In too many cases such statements are deserved. However, the fact remains that groups are used for problem solving and decision making by almost all organizations. Knowing how to use a group effectively, and when to assign a problem to a group and when to assign it to an individual is the difference between success and chaos. Some guidelines, then, on these two techniques should be useful.

CHOOSING - GROUP OR INDIVIDUAL

There are several factors to consider when deciding whether to assign a particular task to an individual or to a group of people for joint consideration. The nature of the task is first and most important. Certain tasks, such as creative or independent tasks, are best performed by individuals; others that involve integrative functions or goal setting are best done by groups.

Individuals working separately are more creative and effective as generators of ideas and as problem solvers than individuals working together in groups. When the task calls for a creative solution, an individual is better than a group. For example, individuals do better than groups at designing a technical component, or writing a computer program. When seeking a creative outcome, find an expert in the area, rather than assemble a group.

When bits and pieces of information are required to be brought together to produce a solution, such as developing a business strategy, or evaluating a new product, groups can produce better results. This, of course, depends on the ability of the group to work together effectively. One way to insure that they don't is to have people on the team whose jobs are, for the most part independent of one another.

Research has shown that when people participate in the decision making process, they have more commitment to that decision. They feel more ownership over the outcome. It is particularly important that persons should be involved in determining the goals that are designed to guide their behavior and against which they are to be evaluated. When goal setting is done in relevant groupings of managers and subordinates, morecommitment to individual objectives can be expected.

However, there may be trade-offs associated with obtaining a group commitment. As an energy manager, you may have to decide between a solution that has a greater acceptance and commitment but is of lower quality, and a solution that is more difficult to implement but is of higher quality.

CHARACTERISTICS OF GROUP MEMBERS

The group should have a shared sense of purpose, and its members should be aware of common goals. Their participation should be voluntary if at all possible. Each person should have some expertise to bring to the particular problem under consideration, and should have some stake in the outcome. In addition, they should have a role to play in implementing any decision. From these guidelines, it is obvious that you probably will not wish to convene the same collection of individuals to address every issue.

WORKSHOPS

By using a structured workshop, an energy manager can get valuable input into many of his or her managerial type problems. I recently worked with a large corporation that was in the process of starting an energy management program at corporate level. An individual has been appointed to head the program, and coordinators had been selected, but the structure along with policy and procedures were yet to be worked out.

The effort was kicked off by having a three day educational conference for the coordinators and key management people. Within this three day program, seven workshops were conducted. The idea was to get an input from the participants on each stage of the planning and organizing of the corporate energy management program. Some of the workshop topics were:

- What are the barriers to the energy management program.
- What is the role of the energy coordinator.
- Where should the program be five years from now.
- What techniques can be used to motivate employees to save energy.
- What energy conservation opportunities can we implement immediately.

The information obtained by this process was valuable because it represented the thinking of the people to be involved in the program. It also provided motivation to the group by allowing them to become involved in the formative stages.

WORKSHOP PROCEDURE

If the output from a workshop activity is to have meaning and value, the workshop must be conducted in a structured manner.

Otherwise, it may simply represent the thinking af a vocal few rather than a consensus of the group.

People who specialize in group dynamics can list a number of techniques for conducting workshops. Some become so elaborate that the process consumes the objective. You really don't need but one technique, so find one that works for you and stick with it.

The one that I prefer and recommend is a simplied version of the Nominal Group Technique (NGT). The NGT was developed in the late 1960's at the University of Wisconsin by Andre Delbecq and Andreq Van de Ven. Unlike typical interacting groups in which all communication among members takes place with minimal structuring or control, the Nominal Group Technique is one in which individuals work in the presence of others but do not interact verbally except at specified times. Written output is generated by each participant and is sequentially shared and listed for all members to see. It is a structured meeting that attempts to provide an orderly mechanism for obtaining qualitative information from groups who are familiar with a particular problem.

A step by step description of the process is as follows:

1. Problem definition

The problem should be defined so well that there is no misunderstanding by the group as to what their task is. You must give some thought to this or else the whole effort can be wasted. The problem should be written out and placed in plain view of the group throughout the workshop activity. It should also be discussed initially to be sure everyone understands.

A good workshop problem to start out with for a learning experience as well as information is "What are the barriers to our energy management program." The top three answers are usually (1 lack of top management support, (2 lack of funding, and (3 lack of trained personnel. You may then wish to have a workshop activity in which each of these topics is the subject for discussion.

2. Grouping

Specialists will tell you that the optimum group size for using the NGT is seven. Five or six will also give satisfactory results. The objective is to not get too small or too large a group. If it is a large audience that you are working with, break them down into these sizes.

The group should then elect someone to be the recording secretary, or perhaps we should call this position the executive secretary because this person should have the responsibility for moving the group along on a time schedule, keeping them within the rules, recording the output, and often making the presentation of the group's findings.

3. Silent Generation of Ideas

After making sure that each person understands the problem, the next step is for each to silently and independently write as many answers to the problem as he or she can generate within the time allotted. And there should be a time limit - generally fifteen minutes is adequate.

4. Round-Robin Listing

At this step, the executive secretary goes around and around the group listing one idea from each until all ideas have been presented. The use of a pad and easel is recommended so all can see the list as it is being generated.

It is important that the ideas not be discussed or rewritten at this stage. They are simply to be recorded verbatim as they are given. A discussion at this stage can impair the free flow of ideas.

This is where most groups get hung up, so the recording secretary must remind them of the rules often.

5. Discussion

After all ideas have been listed, the secretary can then lead the group in a discussion of each idea for the purpose of clarifiction, elaboration, evaluation and combining. Each item is discussed sequentially and no item is eliminated from the list. Like items may be combined. A new list may be generated at this point to reflect this clarification.

6. Ranking

Each person, without interacting with others, is asked to select the five most important items and rank them numerically. The first choice will receive 5 points, the second 4 points, the third 3 points, the fourth 2 points, and the fifth 1 point.

The total number of points received for each idea will determine the first choice of the group. Continue until five or ten ranked selections have been made. These should be written down in rank orded and preserved for the individual responsible for the overall program. An oral presentation should be made by each group if there are two or more groups involved in the exercise.

If there are two or more groups, and you want the consensus of all, you may wish to compile another list made up of the top choices of each group, then allow the whole group to rank them - using the same ranking process.

Chapter 9

THE ECONOMIC EVALUATION

Lack of funds - how many times have you heard this excuse for not implementing an energy project? How many times have you used it? In workshop activities that I have conducted with groups in energy management programs, lack of funds is a recurring answer to the question "What are the barriers to your energy management program." It is either the number one answer or the number two with a great majority of the groups.

In some cases, a lack of funds may be a legitimate answer, but I would venture to say that it would apply to a very small percentage of those who give that answer. With good energy projects, there are simply too many ways in which funding can be accomplished - either internal or external.

The real problem may be that the economic evaluation is not properly done and presented to the decision makers. This is a skill that too many of the technical people involved in managing energy have never acquired. It is one that is needed if you are going to do a creditable job. It can give you a distinct edge over those within your organization who don't have it, and are competing with you for funds.

It is not necessary that you become an expert in economic analysis. That alone is usually a full time job within most organizations - dealing with the tax structure, and the internal guidelines. However, it is necessary that you have a comprehensive knowledge of the process, so that you know if input figures are correct, and which economic evaluation method is best suited for the situation. Your accounting people may be using energy cost figures that are obsolete and could be killing your projects without your knowing it. Cost escalation may not be according to the latest best guesses.

You are in a better position to know these numbers than anyone in your organization, and if these kinds of numbers didn't come from you, they probably need reviewing. This is a fact that many energy managers overlook. It is something that could be very well incorporated into your energy policy.

Every organization has some method of making economic decisions. It varies from a hip pocket, gut feel, to a very sophisticated computer analysis. If you are working with a group at the lower end of the scale, there may be frustrations, but it is generally easier to sell them on a project that does have a good economic analysis. An energy manager that does his or her homework in such a system, presents good sound energy proposals with a good economic analysis is practically assured a lions share of available funds.

The purpose of this chapter on economic analysis is not to attempt in one short chapter to give you all the necessary skills. Instead, I hope to give you enough of the basic fundamentals to whet your appetite for further study. The references provided at the end of this chapter are a good place to start.

Most of this chapter will be devoted to life cycle cost analysis, because that is the fundamental process from which most all economic analyses are made. Before jumping into the details of the economic evaluation process, it might be good to first consider the various methods that are now available for financing energy projects.

FINANCING SCHEMES

Because the return on investment has been so good where energy is concerned, - and continues to be - the financial people have become attracted to the field and have developed several schemes in which both the user and the investor have been very happy.

This is unique to energy projects. No other area has attracted this much financial interest. This tells you something about the potential that exists in energy projects.

There are five basic financial plans currently in use:
- Funding by using organization
- Shared savings financing
- Energy services
- Leasing
- Insurance for guaranteeing energy savings

<u>Funding by Using Organization</u> - Funding of energy projects is most commonly done by the using organization, using one of the methods described below.

One of the primary sources of funding for private firms is *retained earnings*, which is money remaining after taxes and after any cash dividends to stockholders. Using retained earnings to finance a project is referred to as *internal financing*.

Another method frequently used is *debt financing*, in which loans are obtained directly from a bank or insurance company to finance the project.

A third method is called *equity financing*. This involves selling of shares of stock to raise the investment funds.

<u>Shared Savings Financing</u> - This has been a very successful concept that was spawned from the great potential savings available from energy projects. It has been very attractive for those organizations that for various reasons lack the capital for energy conservation projects. It allows them to proceed with investments which they might not be able to undertake for several years using

their own resources. In addition, with the shared savings plan, they receive an immediate positive cash flow.

There are three identifiable groups involved in shared savings plans - the user, or customer, the technical group, and the financial investor. In the earlier years, these groups were three separate organizations, but as they grew, many of the technical groups began to do their own financing, and in fact, offer a turn key operation.

The project starts out with an engineering study that has two basic purposes : to establish the present baseline of energy usage, and determine the potential savings from the proposed energy project. There are many factors to be considered in establishing the baseline, such as degree days, production, new construction, energy conservation by the user, etc. This is sometimes the most difficult area in which to arrive at agreements, but as experience has grown with the shared savings companies, it is becoming more of a science and less of an art.

This engineering study is usually at no cost to the user - if they then subscribe to the program. Otherwise, there is usually some cost associated.

The financial study, which is the next phase, not only determines the amount of money needed to finance the project, but also checks out the financial status of the user. It wouldn't be financially smart to enter into a ten year contract with a company on the verge of bankruptcy.

The contract typically has the shared savings company responsible for the initial energy audit, design and fabrication of a system, all system components, installation, maintenance, and service. The customer makes no capital investment. The user agrees to pay a percentage of the savings generated by the project - usually 50% - for a period of time, which ranges from 5 to 10 years. Tax benefits and depreciation usually go to the supplier.

Typical projects that have been financed by the shared savings concept are:
- Heat recovery
- Recuperators
- New boilers
- Fuel substitutions
- Cogeneration
- Relamping
- Energy Management Systems

<u>Energy Services</u> - I am not sure of the origin of the concept of energy services, but France has many more companies involved in it than does the United States, although the number is growing here.

An energy service company will basically contract with a user to provide given levels of heating, cooling, lighting, and equipment use. They then assume responsibility for the utility bill payment, and make their profit on the spread between its costs and its fees for providing the services - which are less than the user was paying initially.

They will make capital improvements at no cost to the user. The contract may also include other services such as audits, consulting, maintenance, and management.

The process, like shared savings contracts, starts with an engineering study that looks at the energy history of the user, cost reduction opportunities, and will usually evaluate the capability of the staff that operates energy using equipment.

The contract makes a commitment to save a specified percentage for the user. This is adjusted for weather variations.

Energy service contracts are most often employed by apartment house owners, or other commercial structures rather than industrial facilities.

Leasing - One method that a user can employ to use energy savings equipment without incurring large up-front costs is to lease it. A lease is a means to convey the right to use a piece of property without conveying ownership.

Monthly base payments can be made less than energy savings to provides a positive cash flow.

The lessor claims all tax credits and depreciation. Recent tax laws have reduced both tax credits and allowed depreciation, making leasing less attractive to lessors.

People who are involved in leasing equipment include:
- Equipment manufacturers
- Leasing companies
- Corporations searching for tax benefits
- Individual investors

There are two basic types of leases - a true lease, and a finance lease.

A true lease has a term that is shorter than the useful life of the equipment, and the lessee cannot specify terms of a purchase option other than at fair market value of the equipment at the end of the lease period.

A finance lease is really an installment or conditional sales contract. It amortizes the full value of the equipment over the term of the lease. The purchase option can be set at a nominal amount.

Insurance for Guaranteeing Energy Savings - This is a financing scheme that is promoted by equipment manufacturers and contractors, and is a good way for an energy manager to minimize the risk on an investment. There is an added cost of $6 to $10 for each $1000 of project cost for this insurance.

It may be calculated annually over the life of the lease if leased equipment is used. For purchased equipment, the contract may be written for the payback time.

There are some distinct advantages for using some of the financial schemes described above. For one, it can free up capital that may provide a greater return on investment than that available for the energy project. Or if the hurdle rate is high because of good returns, many energy projects that would not qualify for internal funding may be funded through one of the above financing schemes.

There can be an immediate positive cash flow that would not be available through the normal project process, at least until the end of the payback time.

The transfer of technological risk from user to vendor can be a tremendous advantage, particularly to those organizations without much depth in technology.

The public sector, consisting of municipal governments, schools, hospitals, colleges and universities, etc., is faced with dwindling federal support, budget-conscious constituencies, rising costs, aging infrastructures, and many social issues are particularly pressed for capital for energy projects. In addition, they have been slow to take advantage of private sector financing options available.

Recognizing this, the Department of Energy funded the Positive Cash Flow Financing project to develop management criteria for use by the public sector to assist them in selecting the most advantageous financing options from among those available to them. The report is available from DOE.

LIFE CYCLE COSTING

Life cycle costing (LCC) is a method of determining the total cost of ownership over the life of the asset. The life cycle cost of an automobile, for example would include: initial down payment, monthly payments including interest, maintenance, fuel, insurance, titling, and resale value at the time of disposal. If you do this, you will probably never buy another car.

The cost of each of these items occurs at different times, so by just using those numbers in their present form, it would be very difficult, say to compare two different cars that you may be considering, perhaps a luxury car with high trade in versus an economy car with low trade in but good fuel economy.

That's where life cycle costing comes in. By using it, you can convert all those costs into one equivalent cost, either present value or equivalent annual cost.

If time is of value to you, you probably would not go through such an exercise for the purpose of a lower value item such as a fishing rod and reel, where you are more concerned about quality, matching it to your specific need, and getting it by your spouse.

The point is, life cycle costing analysis is expensive, and the implementation of it should be weighed against the value of the project being considered.

In most cases, it is used for comparison of two projects that have similar objectives, but a different mix of costs, so that comparison of their financial value would be difficult without the ability to convert both to one equivalent cost.

One of the major factors that provides job security for energy managers is the purchase of equipment on the lowest initial cost. Companies will construct a new building, then ask the energy manager to make it energy efficient. Sometimes after a proper LCC analysis has been made, and the project implemented, lowest initial bids on the next go around can delete the whole project. This has happened to several energy managers who did a proper study of lighting, selected an energy efficient lamp that was slightly higher in initial cost, then had the purchasing agent replace them with low cost, higher wattage lamps on the next bid. Educating key people in your organization on your energy management goals and procedures is the only hope for that kind of problem.

The major concepts of life cycle costing are the time value of money and the concept of equivalence. These two concepts plus the basic interest formulas used for determining equivalency are the major components of the life cycle costing process.

TIME VALUE OF MONEY - Time and money have an integral relationship. Because of inflation and interest, the value of money changes with time. If $100 is simply put in the mattress and held for a year, its purchasing power has declined. If it is put into a savings account to draw interest, its value will increase. If it is put into a checking account that does not pay interest, then the bank will invest it and make money from it. Money, then is not static in its value. It is either gaining or loosing.

If the $100 that is placed in a savings account to draw interest for a year, then the principal and the interest are left for another year, interest is also paid on the interest from the first year. This is called **compounding**. Figure 9.1 demonstrates compounding in which the $100 is invested for five years at 5% interest. At the end of the fifth year the initial investment is worth $127.76.

CONCEPT OF EQUIVALENCE - We cannot add numerically the initial cost and subsequent costs of a project without discounting the subsequent costs to a present value. Discounting is the process of making all costs equivalent at one period of time,

Figure 9.1

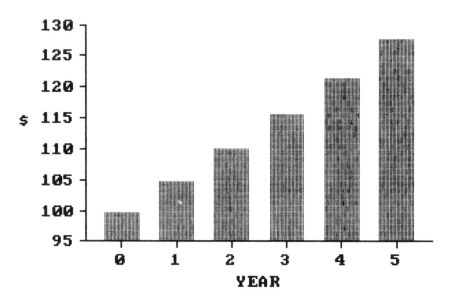

THE ECONOMIC EVALUATION 107

usually the present, although the process allows equivalency to be determined at any time in the future.

Too often decision makers merely add future costs to present costs without regard to the fact that capital has a time value. We are either paying interest to use it, or gaining interest from investing it, and this must somehow be factored in.

<u>BASIC INTEREST FORMULAS</u> - There are six basic formulas that comprise the tools of life cycle analysis. These will be given and illustrated by examples, but first let's define the symbols that are used, then illustrate a method for diagraming the components that will greatly aid in the understanding. The symbols are:

i = **interest rate per period**
n = **number of interest periods - usually in years**
P = **present worth or present value**
F = **future worth or future value**
A = **uniform sum of money in each time period**

The interest rate "i" in most tables is the rate per year, and "n", the number of interest periods, is usually given in years. Don't forget this if you are trying to determine monthly payments.

"P", the present worth is the value of dollars at the present instant in time.

"F", the future worth is the value if we took a present dollar value and projected its worth for some specified time into the future. If it is money that is earning interest, the future value would, of course, be greater. If, on the other hand, it is being eroded by inflation, it would be less.

"A" represents the annual payment in our calculations.

We can, through the discounting process, equate all numbers to a "P" value - present worth - or an "A" value - annual payment.

Figure 9.2

CASH FLOW DIAGRAMS

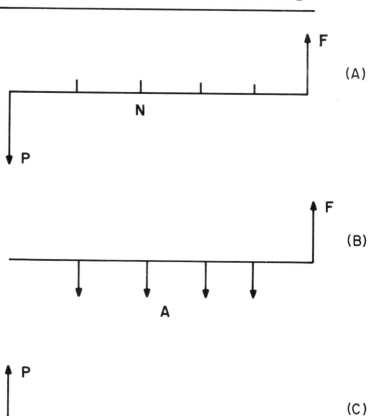

THE ECONOMIC EVALUATION 109

If we do this for two projects under consideration, then we can compare apples and apples.

A cash flow diagram of the above terms is shown in figure 9.2. A downward direction indicates cash flow out, and an upward arrow indicates cash flow in. This visual method of displaying the economics of a project not only helps in the understanding, but is also useful in selling the project to others.

Figure 9A graphically represents an initial investment "P for "n" years that will yield a future sum of "F".

9B shows annual payments "A" being invested to give a future return "F".

9C depicts an initial receipt of money "P" such as a loan, then annual payments "A".

It is important to note the equivalency of numbers represented by the diagram. For example, in Figure 9A, "P" and "F" are not numerically equal, but because of the time value of money, they are equivalent. The same is true of the other two diagrams. The annual payments "A" are equivalent to the future value "F" in 9B and to the present value "P" in 9C.

<u>Single Payment Compound Amount Factor</u> - If we wish to determine the future value of a present sum of money invested at a given interest rate for one period, the following formula is used:

$$F = (P + iP) = P(1 + i)$$

If we re-invest the entire amount for a second period, the amount due at the end of the second period would be:

$$F = P(1 + i)i = P(1 = i)2$$

If we re-invest for n periods, the value of the future sum would be:

$F = P(1 + i)^n$ Formula 9.1

If we divide both sides by P we get:

$F/P = (1 + i)^n$

Then:

$F = P * F/P$

F/P is called the **Single Payment Compound Amount Factor (SCA)** and is shown in the first column of the interest tables at the back of this chapter.

Example: Suppose you wished to know how much money you would have at the end of five years if you invested $10,000 at 12% for that period.

The diagram would look like this:

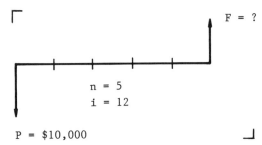

The formula would be: $F = P * F/P$

Looking in the 12% interest table, in Appendix F under the F/P factor, move down to the 5th period. This factor is 1.7623

$F = \$10,000 * 1.7623$

$F = \$17,623$

Single Payment Present Worth Factor-This factor is used to determine the amount of present value money "P" must be invested at a given interest rate and for some period of time to produce a desired amount of future value "F".

If $F = P(1 + i)^n$ then;

$P = F/(1/1 + i)^n$ Formula 9.2

Dividing both sides by F we get:

$P/F = 1/(1 = i)^n$

Then:

$P = F * P/F$

P/F is called the Single Payment Present Worth Factor (SPW), and is given in the second column of the interest tables.

Example: How much money must be invested for 10 years at 12% to have a future sum of $20,000?

The diagram would look like this:

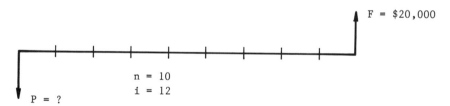

F = $20,000

n = 10
i = 12

P = ?

The formula would be:

$P = F * P/F$

Looking at the 12% interest table under the P/F factor, move down to the 10th period where:

P/F = .3220

Then:

P = $20,000 * .3220

P = $ 6440

<u>Capital Recovery Factor</u> - The amount of periodic payments "A" required to equal a capital investment "P" can be determined by the following equation:

A = P * i(1 + i)^n/(1 + i)^n -1 formula 9.3

Dividing each side by P gives:

A/P = i(1 + i)^n/(1 + i)^n - 1

A/P is called the **Capital recovery factor (UCR)** and is shown in the third column of the interest tables.

Example: What are the annual payments on $15,000 borrowed for 8 years at 15% interest?

This diagram looks like this:

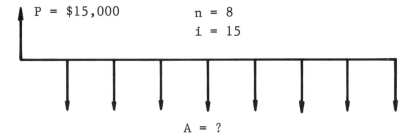

In the 15% interest table, under the A/P factor at the 8th year, find: A/P = .22285

Then:

A = P * A/P

A = $15,000 * .22285

A = $3,342

<u>Uniform Series Present Worth Factor</u> - This model is the inverse of the previous one above and can be used for solving the problem of "How much do we need to have on deposit in an interest earning account if we wish to withdraw a given amount at the end of each period".

The formula for this is:

$$P = A * ((1 + i)^n - 1/i(1 + i)^n) \quad \text{Formula 9.4}$$

Dividing both sides by P gives the A/P factor which is called the Uniform Series Present Worth Factor (UPW), shown in column four of the interest tables.

Example: How much can borrowed at 15% interest for 10 years if payments are to be $1000 annually?

Here is the diagram for this example:

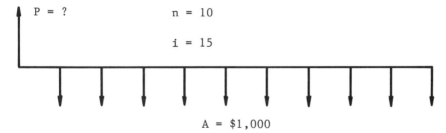

The formula is:

P = A * P/A

The P/A factor from the 15% interest table at 10 years is 5.019.

P = $1000 * 5.019

P = $5,019

<u>Sinking fund factor</u> - This formula will solve the problem "How much money do we need to deposit periodically "A" to have a future sum "F" of a given amount at the end of time n.

The formula for this is:

$$A = F*i/[(1 + i)^n - 1] \quad \text{Formula 9.5}$$

Dividing both sides by F gives the A/F factor, which is called the Sinking fund Factor (USF), and is given in the interest tables.

Example: What annual sum must be invested at 10% for 5 years to have a future sum of $10,000?

The diagram is:

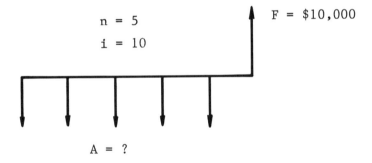

The formula is

A = F * A/F

The A/F factor in the 10% interest table under A/F at 5 years is .16380.

A = $10,000 * .16380

A = $1638

Uniform Series Compound Amount Factor- In this model we can determine the amount of a future sum "F" if we know the periodic payments plus interest rate "i" and the number of periods "n".

The formula for this is:

$$F = A * [(1 + i)^n - 1]/i \quad \text{Formula 9.6}$$

Dividing both sides by A gives F/A, the Uniform Series Compound Amount Factor (UCA).

Example: How much money will accrue if $7200 is invested annually for 12 years at 10%?

The formula is:

$$F = A * F/A$$

The F/A factor from the 10% interest table under F/A at 12 years is 21.384.

F = $7200 * 21.384

F = $153,964

<u>Factor Selection Procedure</u> - Using the above formulas may be easy, but determining which to use may present a problem. In order to simplify the process, the following procedure may help in selecting the correct factor.

First, put the symbols in two rows, one above the other as below:

P A F (known)

P A F (unknown)

The top represents the known values, and the bottom line represents the unknown. From information you have and desire, simply circle one of each line, and you have the correct factor.

For example if you want to determine the annual payments on a loan, then "P" is the known, and "A" is the unknown, so circle P on top and A on bottom for a factor of P/A.

Table 9.1 is a summary of the above.

Summary of Factors

Factor	Expression	Name
$\frac{F}{P}$	$(1 + i)^n$	Single Payment Compound Amount Factor
$\frac{P}{F}$	$\dfrac{1}{(1 + i)^n}$	Single Payment Present Worth Factor
$\frac{A}{F}$	$\dfrac{i}{(1 + i)^{n-1}}$	Sinking fund factor
$\frac{A}{P}$	$\dfrac{i(1 + i)^n}{(1 + i)^n - 1}$ $= \{\dfrac{i}{(1 + i)^n - 1} + i\}$	Capital recovery factor
$\frac{F}{A}$	$\dfrac{(1 + i)^n - 1}{i}$	Uniform series Compound amount factor
$\frac{P}{A}$	$\dfrac{(1 + i)^n - 1}{i(1 + i)^n}$	Uniform series Present worth factor

COSTS

Determining all costs that should be considered in a life cycle cost analysis may be done by developing your own all-inclusive list of those things that should be considered.

Some broad categories are listed below for your assistance in making such a list:

- Initial Investment - may include cost of study, design, acquisition, installation.
- Operating costs - staff, energy, insurance, maintenance.
- Salvage or disposal
- Cost of money - interest.
- Depreciation
- Taxes
- Cost escalation factors - interest, energy, taxes.

SAMPLE PROBLEM

Working through the following sample problem should help you grasp the concepts explained in the preceding material.

Given:

- First cost = $1,000,000
- Life of equipment = 20 years
- Annual fixed cost = $20,000
- Salvage = $50,000
- Periodic expenditure in year 5, 10, & 15 = $40,000
- Cost of capital = 10%

118 MANAGING ENERGY RESOURCES IN TIMES OF DYNAMIC CHANGE

Find:
- Equivalent present worth
- Equivalent annual cost

#1 Solution: Equivalent present worth

First draw the cash flow diagram. This is shown in figure 9.3.

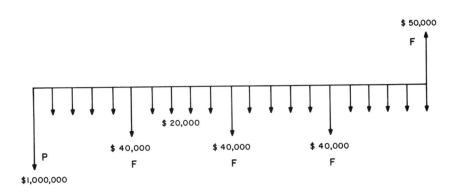

Figure 9.3

Figure 9.3

- The first cost is a present worth, and is a cash outflow, so we give it a negative value.

$P_1 = -\$1,000,000$

- Annual cost discounted to present worth

$P = A * P/A$

$P/A\ (i=10\%, n=20) = 8.514$

$A = \$20,000$

$P = \$20,000 * 8.515$

$P_2 = -\$1,702,800$

- Salvage cost discounted to present worth

$P = F * P/F$

$P/F\ (i=10\%, n=20) = .1486$

$F = \$50,000$

$P = \$50,000 * .1486$

$P_3 = +\$7430$

- Periodic expenditure discounted to present worth

$P = F * P/F$

$P/F\ (i=10\%, n=5) = .6209$

$P/F\ (i=10\%, n=10) = .3855$

$P/F\ (i=10\%\ n=15) = .2394$

$F = \$40,000$

$P = \$40,000 * (.6209 + .3855 + .2394)$

$P4 = -\$49,832$

Totals + $7430 - $2,752,632

Total Present Worth Pt = $2,745,202

#2 Solution: equivalent annual cost.

- First cost discounted to annul cost

A = P * A/P

A/P (i=10%, n=20) = .11746

P = $1000,000

A = $1,000,000 * .11746

A1 = - $117,460

- Annual fixed cost

A2 = - $200,000

- Salvage discounted to annual cost

A = F * A/F

F = $50,000

A/F (i=10%, n=20) = .01746

A = $50,000 * .01746

A3 = + 873

- Periodic expenditure discounted to annual cost. Since three different n values are involved, we must first discount back to a present value, then find the equivalent annual value. From the previous solution, we have:

P = $49,832

A = P * A/P

A/P (i=10%, n=20) = .11746

A = $49,832 * .11746

A4 = - $5,853

0 = + 873 - $323,313

Total equivalent annual cost: - $322,440

#3 Solution equivalent annual cost:

Since we determined the equivalent worth value for the whole project in solution #1, we can simply convert this value to the equivalent annual cost.
- Equiv. present cost = $2,745,202
- A = P * A/P

A/P (i=10%, n=20) = .11746

A = $2,745,202 * .11746

A = $322,451

If you were trying to compare two projects from an economic aspect, you can readily see the advantages of discounting all the costs to either an annual cost or a present worth. You have one number from each project to compare.

PAYBACK

Payback is the length of time required for invested capital to be recovered from net cash flows.

$$\text{Payback} = \frac{\text{net investment}}{\text{net annual cash flow}}$$

If an investment of $20,000 provides after-tax savings of $4,000 per year, the simple payback is $20,000/ $4,000 = 5 years.

Some organizations in their simple payback calculations include such things as salvage value, depreciation, and taxes. If there is a set procedure established by your organization, you as energy manager should be aware of it so that your projects are evaluated the same as others.

Simple payback does not take into consideration the time value of money, but it is an inexpensive way of making an initial determination if a project is in the ball park. Some organizations have an established hurdle rate for payback. Quite a few have two years as a maximum payback period for projects to even be considered. This hurdle rate generally depends upon the economic health of the organization, and can vary from time to time.

If projects are below a given capital investment level, a payback calculation may be all that is necessary in the way of an economic evaluation. If your organization does not have an established hurdle rate, or a maximum level for just a payback analysis, you should consider establishing one. Having this understanding with those who control the funding can save you a major sales effort with each project you try to promote.

RATE OF RETURN

The rate of return is simply the reciprocal of payback.

$$\text{Rate of return} = \frac{\text{net annual return}}{\text{investment}}$$

From the above example, the rate of return is:

$$\$4,000/\$20,000 = 20\%.$$

INTERNAL RATE OF RETURN

This is perhaps the most common economic evaluation method employed for projects involving a substantial capital investment.

The internal rate of return is the discount rate at which the net life cycle savings equals zero. It is used to compare projects which have a net positive cash flow after taxes. In this method, no interest rate "i" is selected, instead it is calculated. In some cases, it is possible to solve for this rate in closed form. More often, we must solve for it iteratively by selecting different values of "i".

The procedure is as follows:

(1) Calculate the cash flows both receipts and disbursements for all years.

(2) Use a trial discount rate "i" and calculate the present worth of each cash flow. Consider each to be a future value "F" when discounting.

(3) Sum the discounted cash flows.

(4) If the sum is zero, the rate of return "i" selected is the proper rate of return. If not, select a new rate "i" and repeat steps (2) through (4).

Chart 9.2 gives a sample evaluation using the internal rate of return.

		YEAR										
		0	1	2	3	4	5	6	7	8	9	10
INITIAL INVESTMENT	A	(1000)										
ENERGY SAVINGS	B		500	535	572	613	655	701	750	803	859	919
COSTS	C		(100)	(104)	(108)	(112)	(117)	(122)	(127)	(132)	(137)	(142)
NET SAVINGS (B + C)	D		400	431	464	500	538	580	624	671	722	777
NET SAVINGS X(1-TAX RATE)	E		200	216	232	250	269	290	312	336	361	399
DEPRECIATION	F		143	209	200	200	200					
DEPRECIATION X TAX RATE	G		71	105	100	100	100					
INVESTMENT TAX CREDIT	H		100									
ENERGY TAX CREDIT	I											
CASH FLOW A+E+G+H+I	J	(1000)	371	320	332	350	369	290	312	336	361	389

INTERNAL RATE OF RETURN - 32.2 %

COMPUTER SOFTWARE PROGRAMS

There are several computer software programs such as EN-VEST developed by Alliance to Save Energy that will perform the calculations described in this chapter. However, before they will give you a complete economic evaluation, you must input numbers for future energy costs. This is not an easy task in times of dynamic change, but the computer accepts only numbers so there is no way around making these predictions. The next chapter doesn't make these predictions for you, but will give you some sources of information and guidelines for making your own.

Bibliography

Brown, Robert J. and *Rudolph R. Yanuck*, Introduction to Life Cycle Costing, The Fairmont Press, inc., Atlanta, 1985

Turner, Wayne C., Energy Management Handbook, John Wiley & Sons, New York, 1982.

Thumann, Albert, Plant engineers and Managers guide to Energy Conservation, Van Nostrand Reinhold Company, 1983.

Dame, Richard E., "Economics of Energy Usage", Energy Management Manual, Virginia Polytechnic Institute and State University, 1986.

Chapter 10

FUTURE ENERGY COSTS AND AVAILABILITY

I am not going to try to predict what future energy costs and availability will be. I will leave that to people more courageous than I. Instead I will simply provide you with references that can supply such projections.

BASIC FACTS

There are some basic facts that there does seem to be a consensus on. First, in spite of much publicity about alternate sources of energy, oil and natural gas will continue to be the prime sources of energy through the end of the century.

Second, as we begin to approach the worlds capacity to produce oil, the price will begin a gradual rise. There may be other influencing factors that can skew this trend temporarily, but it will eventually override them.

Twenty million barrels of oil per day is the pivotal point with OPEC. Below this point there is a problem in having members stay within their allotment. Therefore, pricing is more competitive. Above the twenty million, this is less bothersome. Conservation and new discoveries may keep production below this level into the 1990's.

The oil reserves of OPEC are so much greater than the rest of the world that it is just a matter of time until they will be asked to produce at capacity to meet world needs. This fact too often is overlooked in our dealings with the OPEC countries.

Most projections of cost and availability are so similar that I sometimes wonder...is there, is there not one Guru somewhere on a mountain that all go to. The consensus is that there are three scenarios, and we are free to pick either.

The low scenario is that oil prices will increase moderately until the year two thousand, rising to around $28 per barrel. The middle scenario is that they will rise to $40 - $50 per barrel and the high one is that it will go to $60 per barrel by 2000.

There is still another one which is probably more realistic. It predicts a cyclic swing in oil prices. There is a tremendous storage capacity now available throughout the world in the form of tankers, tanks, and even underground caverns. This has a controlling effect on the world market and tends to create this cyclic trend by buying when the price is low, then use stored capacity when it is high.

DEVELOPING YOUR OWN CRYSTAL BALL

Many large firms have access to consultants who constantly analyze energy price and availability, and report this to the organization periodically. In the absence of such a service, you should develop your own sources. I will discuss some potential sources below.

The Wall Street Journal provides a daily listing of crude oil prices. It also occasionally contains in-depth articles relating to energy.

Energy User News periodically has a survey of what energy managers think prices are going to do in the future. They also have a "Price Watch" section listing current energy prices.

The National Bureau of Standards issues projected energy price escalation rates which are to be used for performing life-cycle cost analysis of design alternatives for either new or existing Federal Government buildings.

The Department of Energy issues several reports dealing with cost and availability projections, and energy policy which contains projections.

The World Bank does a complete energy analysis on any country seeking a loan. Their information on a global basis is very good.

Several of the oil companies produce reports on cost and availability. Their reports contain very good graphs and charts that can be valuable in preparing a presentation. Conoco and Chevron are two that make very good reports available at no charge.

Addresses for the above sources are given below:

Wall Street Journal

Wall Street Journal, 200 Liberty Street, New York, NY 10281
Telephone: (212) 416-2000

National Bureau of Standards Handbook 135

"Life Cycle Cost Manual for the Federal Energy Management Program". Available from the Superintendent of Documents, United States Government Printing Office, Washington, DC 20420 for $11.00. Expanded economic analysis data is contained in NBS Pamphlet NBSIR 85-3273, "Energy Prices and Discount Factors For Life cycle Cost Analysis." Available from the National Technical Information Service, Springfield, VA 22161 for $11.95.

Department of Energy

Office of Policy, Planning, and Analysis Washington, DC 20585 (Public Affairs Office: 202-586-6827)

World Bank

Economic Analysis and Projections Department 1818 H. Street, N.W. Washington, D.C. 20433

Conoco Inc.

Public Relations Department 1007 Market Street Wilmington, DE 19898

Chevron Corporation

Economics Department 225 Bush Street San Francisco, CA 94120-7137

Chapter 11

PRESENTING YOUR IDEAS

I have conducted many workshops, using the nominal group technique, in various organizations such as corporations, school systems, and local governments. The first problem I like to give such groups is "What are the barriers to your energy management program."

The ranked order of perceived barriers that emerge from these groups follows an almost identical pattern. The one that is almost always in the first place is "Lack of top management support."

The participants may truly feel that top management does not support their efforts, but it really indicates that they may not have the proper skills to present their case. Very few top managers got to their position without the ability to recognize and act on good projects. Making them recognize good energy projects, then, is a skill that energy managers need.

Being able to present your ideas in a convincing manner is crucial to professional development in any field. It is of particular importance in energy management for two reasons: First, top management probably knows less about managing energy than any other responsibility they may have. Second, it is a technical subject that must often be presented in non-technical terms.

A young engineer attending one of the first sessions of our Energy Management Diploma Program had just been asked by his top management to put together an energy management program for the corporation. He had several weeks to develop a plan, then was to present it to the board of directors for approval. He had fifteen minutes scheduled on the agenda in which to present his case. His whole program hinged not only on his plan, but how well he presented it.

Your situation may not be this dramatic, but the forward progress and growth of your energy management program will certainly hinge on your ability to present yourself and your ideas in a professional manner.

Lewis Powell, in his book <u>Executive Speaking</u> states that "Those who speak articulately with confidence create subconscious conviction."

If you look at any major project or innovative thrust, you will almost always find one individual who is a mover and shaker behind the effort. Very seldom is it a group that causes such projects to attain and sustain momentum. As energy manager then, the responsibility rests heavily on your shoulders alone to make the program effective. This does not preclude integrating all the planning and the involvement of others as proclaimed in the preceding chapters. It simply means that you must be the spearhead. The fact that you are reading this book indicates that you have the potential and recognize the need for continued learning. So, assign the development of good communication skills a top priority in your self development program.

This chapter does not contain all the information you need to know in order to become an expert communicator, but it is intended to emphasize the importance of good communication, provide some tips that you can implement immediately to enhance your presentations, and provide you with references for further study.

WHO IS YOUR AUDIENCE

The audience may be an individual that has approval authority, a top management group, your technical staff, an energy committee, employees, or a combination of these. In any case, it is important to take a few minutes to define and think about the audience that you will be working with for a specific presentation

because you may wish to use different techniques for each. Some questions you might consider are;
- What is the energy knowledge level of the group?
- How technical should the presentation be?
- How large will the group be?
- How many major points should you try to make?
- Should you emphasize details or give an overview?

FIRST IMPRESSION

Dress to kill, dress to impress, or be yourself - these are decisions we have to make every morning, and we may have already established an impression with those who will make up the audience. My only suggestion is to dress sharp in your every day activities, and then when you come up to give a presentation, don't over dress to the point that you look unnatural. Wear something that makes you feel good about yourself without looking like a kid that has just been grabbed up and dressed to go to Sunday School.

WHAT DO YOU WANT TO SAY

The purpose of a presentation generally is one of two things - to provide a learning experience for someone, or to get approval. In some instances, it may take the first to get the second. But in any case, the purpose should be thought out and defined in very few words. Some pointers on giving a good presentation on energy topics are as follows:
- Have one or two major points that you wish to make, then structure your presentation around them.
- Don't start your introduction with the 1973 oil embargo. This has been beat to death.
- If there are numbers and statistics that comprise a portion of your presentation, put them in graph or chart form so they can be immediately and easily comprehended by the

audience. Don't expect them to remember numbers in order to understand the point you are making.

- Don't go into a lot of technical detail in the presentation. Have the data available so if questions arise, you can pull it out, but not before.

- Create interest by first being dynamic yourself. If you follow a speaker who has been rather dry, or if for any reason the audience appears to be tired or bored, when you come on, have them stand up and stretch. Relate to things that are of interest to the audience and to which they may have first hand knowledge.

- Telling a joke that is unrelated to your presentation is amateurish, and can break the line of thought of your audience. However, tying in some humorous incident can indicate wit, and can do a lot to loosen them up. Be careful how you try to use humor. It can enhance a presentation if done properly and in good taste, but can backfire if not presented properly.

- Now take out your "highliter" pen and mark the following. <u>Don't apologize during your presentation for any weakness</u>. Whether it is a lack of knowledge of some topic on your part, or an equipment failure, don't start with an apology. It indicates a weakness on your part that will remain with the audience throughout your whole presentation. In the seminars that I present, I have on occasion apologized for something, such as cutting some subject short because of time constraints. When the evaluations come in from the participants, they reflect back -not enough time for the subject - or whatever I may have apologized for.

I'm sure you have seen a speaker start off by saying something like "If I appear a little nervous, it's because I'm not used to speaking before a group." You then spend the rest of the time watching for signs of the speakers nervousness.

TIME AND TIDE WAIT FOR NO SPEAKER

The part of your presentation given after your allotted time has elapsed usually falls on ears more attuned to listening for "Let's take a break." You must learn to use time for your best interests. It is an asset or a liability, with not much in between. If your major punch is in the last part of your speech and you wind up crowding it into a few remaining seconds in order to stay on time, you loose. On the other hand, if you go over allotted time to make your big pitch, you also loose.

Even if the speaker just before you went way over his time, the audience will still hold you at fault if you go over yours. Not fair maybe, but reality. Therefore, you need some expanders and some contractors in your presentation that will allow you to finish right on the nose. That's professionalism, and will win you everlasting gratitude from a nervous chairman trying to keep everything on schedule.

WONDER WHAT THIS SWITCH DOES

I once was a speaker in an energy seminar conducted for a large corporation in their headquarters building. The lectern had a control panel like a B52 bomber. It would move the lectern up and down, dim lights, change volume, and some other functions I wouldn't dare touch. The only problem was, when I walked up to speak was the first time I had seen this electronic marvel. I asked the audience to take a stand up break, then quickly located the critical controls needed for my presentation.

If at all possible, go in the night before, check out the equipment, and get it all set up. If you try to do it the next morning, there may be a problem that cannot be hurriedly resolved - and there is always some early bird who wants to discuss something very important with you.

Know how to turn on the projector -both slide and overhead. There must be an award for the designer who can best make switches look like anything but a switch, so look out for award winning projectors.

The room arrangement should also be checked. This is usually assigned as a menial task to someone who knows little about making participants comfortable or best viewing arrangement. Chairs are often spaced too close together.

Room temperature is critical and should be checked early enough to allow you to track down some individual who knows the idiosyncrasies of the control system and time to bring the temperature to the proper level.

AUDIO VISUALS - A HELPING HAND

Most presentations can be enhanced by the use of audio visuals since much of our learning takes place through our visual senses. They can also be a crutch that can take the attention away from a nervous speaker and will help eliminate delivery errors.

The process of making audio visuals requires you to better organize the content for easier understanding. We really learn best when we are required to teach, and this is one of the reasons.

Delivery time is usually shorter with the use of audio visuals, and certainly is easier to control. Also, if the presentation is to be repeated at a later date, then less preparation time is required for that next time.

Here is an important thing to remember:

> "A basic objective of audio visuals should be to simplify a presentation by giving one or two main points in a dramatic manner through the use of simple graphs, bar or pie charts, pictures, or a limited number of words or numbers."

Now get out your "highliter" pen again and mark the following: <u>Make your audio visuals look professional</u>. The quality of your audio visuals will reflect how the audience perceives the quality of your overall presentation. This is particularly true if you are following another speaker who does have good audio visuals.

I once conducted a seminar for a large corporation, and used professional quality audio visuals throughout. I was followed by the corporate energy manager who used overheads that he had roughed out by hand. It gave the audience the impression that he didn't care enough about his program to treat it professionally. His top management also noticed this and criticized him for it.

THE BLACKBOARD

Some of the most dynamic presentations I have seen were by speakers who simply walked up on the stage with nothing but a blackboard, and with a flurry of sketches and single words or phrases put on the board as they talked, held the audience right in their hand as they got their points across.

Their points were developed right before your eyes. This takes speaking skill and imagination. You should give it a try sometime, but try to pick your time and topic. I was once asked to be a guest lecturer in an academic class. The topic was Solar Energy, one that I had presented many times with the use of a nice set of slides. As the class began to filter in, the projectionist took the slide tray and inadvertently turned it upside down without the locking ring. There was no way I could sort and reload the tray, so I went to the blackboard and proceeded to give what I thought was my best lecture on Solar Energy. I didn't get a standing ovation but I did get invited back.

Don't overlook the use of colored chalk, and don't feel that you have to have artistic skills. You should perhaps practice writing large, and legibly - and horizontally.

One major advantage of using a blackboard is that you don't have to lower the lighting level. You don't even have to know how to lower the lighting level. Neither do you have to worry about the availability or operation of audio visual equipment.

Anytime you can emphasize a point you wish to make by marking either on a blackboard, a flip chart, or an overhead you give the audience the feeling that the program is not so canned, but is being adapted to their specific interest.

THE FLIP CHART

Using flip charts as audio visuals also has the advantage of not having to lower the lighting level or worrying with audio visual equipment.

Their use should be limited to an audience of not more than 30 to 35 people. Otherwise, they may be too small to be seen.

You can use the flip chart like a blackboard and illustrate as you talk, or have them prepared beforehand. If they prepared beforehand, you may wish to leave blank sheets in between in case you need to write or sketch additional material.

One precaution in the use of flip charts - don't fasten them against a wall and write on them. The ink may come through onto the wall. I have known of cases where a new paint job was required when this was done.

THE OLD RELIABLE OVERHEAD

The overhead projector is most commonly used for audio visuals. The slides are easy to prepare, and operation of the equipment is easy and flexible. Some suggestions that will help make your presentation using overheads visuals more effective are given below.

Because the overhead projector is close to the screen, it is not necessary to turn out the lights. Turning out lights directly over or near the screen may be desirable however.

If you have any control over the screen, select a matte surface rather than beaded. This will allow you to have a wider viewing area. The beaded screens tend to fade out the picture when viewed from the side.

The screen should be of ample size so that everyone in the audience can read all material on the slides. A rule of thumb is to not have the depth of the room exceed 6 times the width of the screen. For example, if the room is 36 feet deep, use a screen that is 6 feet wide as a minimum.

Prevent a keystoned image - that is one in which the bottom is much narrower than the top - by tilting the screen.

Drag out your "highliter" pen again and mark the following. Don't block the screen. This is the major offense committed by speakers using an overhead projector. There is a tendency to work at the projector, and I have seen experienced speakers either block the screen from view or block the projection to the screen. You should either work at the screen with a pointer, or make sure that your position at the projector does not interfere. A knitting needle makes a good pointer if working at the projector. Remember that all movement is amplified when it is projected, so keep the point against the slide to prevent giving the impression you are shaking - even if you are.

Preparing good overhead visuals - First let's look at a couple of things you shouldn't do:

- Don't copy all or a large part of a printed page. Excerpt just the portion you need then make it large enough to be seen by reprinting if necessary.
- Don't copy long columns or figures. It may sometimes be advantageous to show such a format that contains a large number of figures, but then create another visual to work from that contains the specific information you wish to show.

Figure 11-1

HOW TO...
DESIGN OVERHEAD TRANSPARENCIES

	YES	NO
• Horizontal format		
• Visual ideas		
• Single concept		
• Minimum verbiage		

Figure 11-2

HOW TO...
DESIGN OVERHEAD TRANSPARENCIES

- Key words

- Legibility

- Overlays

Figure 11-3

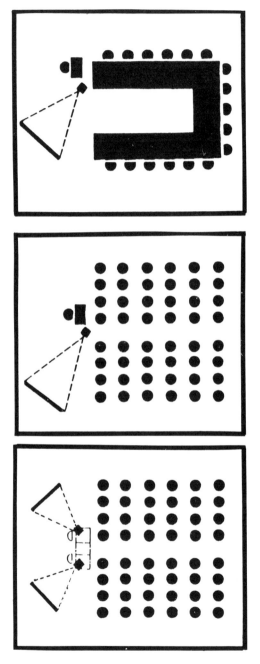

Now, some suggestions for making good visuals:
- Use pictures to supplement words. A presentation that has all words can be enhanced greatly by occasionally injection a picture even if it is a simple sketch.
- Use graphs to present numbers. The mind simply cannot interpret trends and variations nearly as quickly with numbers as with a simple graph.
- Design charts and graphs in a horizontal format.
- A good standard to follow is:

>One idea per visual
>
>Six words per line
>
>Six lines per visual

Figures 11-1 and 11-2 reflect some of these ideas.

<u>Viewing arrangements</u> - Figure 11-3 shows best location of the screen for various seating arrangements. Notice that in all cases, the screen is at an angle to the group.

<u>SLIDES</u> - It has been my observation that a presentation given with the use of slides tends to be more formal. The audience always seems reluctant to interrupt or ask questions. The fact that the lighting level is lower I am sure contributes to this attitude. But I think it is one that should be recognized.

There may be times when you desire to have a more formal approach. If the audience is large and you don't for various reasons want audience participation, then the use of slides can help.

The use of slides does not by any means preclude audience participation. The speaker just needs to have some built in techniques for getting them started by doing such things as asking them questions. People always seem to feel better about a seminar or presentation if they have in someway had an opportunity to provide some input.

The same suggestions for making overhead visuals also applies to slides. Some additional tips for slides are:

- The background should be a light color, but not white. It gives too much glare. Light blue as a background and white letters are always good. If you make slides of a list of topics so that each slide presents a new topic but retains the others, let the new topic have a different color. Magneta is a good color for the used topics when combined with the blue background with white letters, but it is not a good prime color.
- If you need to repeat a slide in your presentation, use duplicate slides, don't turn back.

Slides do provide you with the ability to incorporate more pictures into a presentation - as compared with overheads. I recommend this. If for example you are talking about an energy committee in an organizational structure presentation, include a picture of a group of people in a workshop environment.

<u>COMMON ERRORS</u> Some common errors in visual design that you wish to avoid are:

- Too much information on one visual.
- Words and numbers too small to be read by viewers in the back - or those with less than 20-20 vision.
- Lettering is so fancy it is difficult to read.
- Every visual looks like the last one.

NOW THAT YOU'RE ON YOUR WAY

I have tried in this one chapter to highlight some of the things you should and should not do. I hope this inspires you to pursue the goal of being an accomplished speaker. For practice, I recommend you join the Toastmasters Club. For further reading, I am including a bibliography of material that should prove helpful.

BIBLIOGRAPHY

J. Lewis Powell, Executive Speaking, The bureau of National Affairs, Inc, 1980

William S. Tacey, Business and Professional Speaking, Wm. C. Brown Publishing Company, 1971

Michael Kenny, Presenting Yourself, John Wiley & Sons, Inc.,1982

A. J. MacGregar, Graphics Simplified, University of Toronto Press

Chapter 12

HOW TO KEEP AN ENERGY MANAGEMENT PROGRAM ACTIVE

One of the challenges now facing energy managers is to prevent their program from falling into the category of a fad, and passing from the scene. There are a lot of precedents for this -value analysis, quality circles, management by objectives - just to name a few. These have not entirely passed from the scene, but did have a peak period followed by a sharp decline.

The declining risk of disruption of energy, coupled with a de-emphasis of energy at corporate level have added to the necessity for having a definite strategy to prevent energy management from declining so low in your organization that it would be difficult to pull it back as a viable structure.

In reality, all indicators point to permanency and personal opportunity. But, the energy manager should first convince himself of this, because most good endeavors in any field usually have one champion that continually sparks it.

Once this conviction of permanency is established, a continual rate of progress should be programmed, rather than a burst of effort resulting in burn-out.

With this basis then of permanency and continual rate of progress, a plan that you may wish to consider is one I will call energy networking. By energy networking, I mean integrating the energy management program into every aspect of the organization in a subtle but decisive manner.

Figure 12.1 illustrates the process graphically. You may wish to graph a process specifically for your operation.

Figure 12.1

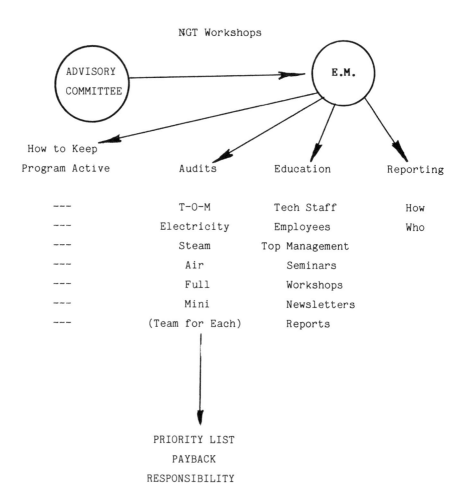

Referring to Figure 12.1, the energy manager is of course the focal point for all activities, but will use various groups in a systematic manner to help with the planning.

The advisory group shown may be one already in existence in your organizational structure, but may operate under a different title, such as energy committee. If they are creative and supportive, use them. If not, you may with to consider setting up a group specifically to help you with this phase of the overall planning.

Use the nominal group technique described in Chapter 8 to develop prioritized lists of ways to keep the program active.

Four suggestions are shown on Figure 12.1. The first is to give the group the problem "How do we keep the program active?" This should generate many good ideas for you to work with.

The second is to develop a list of the kinds of energy audits you wish to conduct. You may have already done it if you have followed suggestions in Chapter 3, Audit Planning. If so, then include it in this planning also. If not, you may use the same advisory group to help with this, or create a new one specifically for this task. A separate group for each of the four suggested areas does have the advantage of involving more people - which is the main objective.

Your educational plan developed in Chapter 4 will provide another list of events that can be integrated in on a prolonged time scale. The list may be further enhanced by using the advisory group for suggestions.

Reporting on a scheduled basis is a tickler that reminds people that the program is active and steaming along.

Once all these lists are complete, the task then becomes one of scheduling them on a timely basis so that there is a small burst of effort going into the program on a regular and systematic basis.

This kind of effort then involving auditing, education, reporting and other things the group may envision can form a network through the organization somewhat like a spider web, so when there is a small activity, it is felt throughout the network.

Chapter 13

TIPS FOR SUCCESS

In observing energy managers that have successful programs, I have identified five specific things that they do which I am convinced plays a critical role in their success. I will describe each below.

GIVE AWAY THE CREDIT

It is very difficult to have a super idea, then have someone else apparently receive most or all of the credit. But, this is what you must accept if you are going to be successful. One sure way of preventing an idea from being implemented is to be too possessive. This is particularly true in managing energy where the support and involvement of everyone in the organization is necessary. If you are creative or if you provide the nourishment for creative ideas, the word will get around without you having to tag each idea with your brand. There is an adage that says "when the water rises, all the boats go up." It's a good one to remember.

Giving credit may sometimes require force feeding. One energy manager related to me the following incident. An operator was using 100 psi steam to clean drums, and was convinced that nothing less would do the job. That night the energy manager had the pressure reduced to half without telling the operator, who the next day proceeded to adequately clean the drums. The energy manager then congratulated him on the energy savings, had his picture placed in the company newsletter, and turned him into an avid energy conservationist.

BE AGGRESSIVE

A good rule to follow is: Don't wait for management to tell you what to do. Make them tell you when to stop. In all likelihood management does not know what you should be doing or how to do it. Energy management is such a new discipline that very few management people have received any training or have previous experience to rely on.

What you may not realize is that you are probably the most knowledgeable person on energy matters in the organization, and as such should take the initiative in moving the organization in the direction it should go with respect to energy.

As the saying goes "Ask for forgiveness, not permission."

USE PROVEN TECHNOLOGY

Don't buy serial #1. Let someone else do all the debugging of new equipment and new technology, or else you may find all your efforts going to try to make something work that you have made a heavy commitment to. I am aware of a firm that, for example, has spent all its energy management effort trying to get a cogeneration system going with a Stirling engine. A tour of their facilities reveals many steam leaks and inoperative steam traps that should have had first priority.

If you want to experiment with new technology, then designate it as a research portion of your overall program and identify it as a high risk project.

GO WITH THE WINNERS

One mistake energy managers often make is to try to keep each subdivision of their organization progressing at the same rate. They usually spent much of their time like a herd bull trying to keep all the cows in a group.

A much more effective way is to devote the time and effort to those that are most receptive to the program. Use them as examples of what can be done. Give them high visibility to top management and to the others involved. Soon those that are dragging will be asking for assistance.

HAVE A PLAN

The best shield against spurious disruptions is to have a plan into which good suggestions can be fitted in proper sequence. Without such a plan, you will be constantly bombarded with suggestions for saving energy by top management, your energy committee, individuals, and in essence, from anyone in the organization - all who will expect you to analysis and report back on their idea.

The preceding chapters have laid out a general plan around which you can develop your own specific plan. Do it!

Chapter 14

UNDERSTANDING AND REDUCING ELECTRICAL COSTS

This chapter is not a part of the planning process as are the other chapters. However, it has been my experience that many energy managers do not have a fundamental understanding of the many variables that make up the cost of electrical energy. Since this understanding is critical to reducing such costs, I felt it would be appropriate to include this chapter in the book.

Even if you have a good knowledge of your electrical costs, a review of these fundamentals will help you better understand the efficiency of electric motors and other energy using devices. For example, why are power factor controls more economical on small motors than large? By understanding the fundamentals of power factors, this will become apparent.

Since it is important that everyone involved in your energy management program have a good basic understanding of electrical costs, you may wish to use some of the material in this chapter to help in this educational effort.

Each electric utility is different. The rate structure, the power generation mix, the reserve capacity, the state regulations under which they operate, the efficiency with which they operate - all are different. Therefore, after you have educated yourself on the basic fundamentals in this chapter, your next step is to talk with your utility and find out about their specific characteristics. But first, the fundamentals.

RATE SCHEDULES

There are three basic measurable components of electrical costs which utilities may include in their rate schedule. They are as follows:

- Energy - expressed in kilowatt-hours or kwh
- Demand - expressed in kilowatts or kw
- Reactive Demand - expressed in reactive kilovars or kvar

In addition two these three, most utilities have a number of other charges that may be included in the overall cost. These include charges such as the following:

- Fuel adjustment clause
- Seasonal rates
- Ratchet clause
- Time of day
- Interruptible service
- Cogeneration

This list is certainly not all inclusive, but is intended to point out typical adjustment type clauses that you may find in utility rates. These cannot be overlooked if you are attempting to calculate your own costs from the electric bill. Fuel adjustment costs for example may be equal to or greater than the cost of the metered kwh.

How many of the above components are included in an electrical bill depends upon the rate schedule under which you are operating. I should point out that it is unusual to find anything but energy - or kilowatt-hrs - being charged in a residential account. There are some experimental programs being conducted to

UNDERSTANDING AND REDUCING ELECTRICAL COSTS

determine the economics of including demand, and a few utilities in the country do have a demand charge for residential customers, but very few. It is important to point this out to people who are interested in saving energy - and money - at home. Otherwise, they may be taking steps at home that will be of no economic benefit to them.

In rate schedules for small general service, energy (kwh) and demand (kw) are usually the only measured charges. A charge for reactive demand (kvar) is generally limited to the large general service rate schedule, and then only if the power factor is low. Power factor will be discussed in following sections of this chapter however because it is necessary to the understanding of motor efficiency.

POWER AND ENERGY - WHAT'S THE DIFFERENCE

It is necessary before progressing further into the technicality of electrical energy to understand the relationship - and difference - between power and energy.

Power may be defined as the rate of expending energy. It is **an instantaneous measurement**. Keep this in mind. It is the product of the instantaneous voltage and current. Therefore, being an instantaneous measurement, it is constantly varying. It varies with the load, and even varies with each cycle of the 60 hertz generation.

In measuring power, then, we are interested primarily in the peak or maximum power. From a designer standpoint, this is needed in order to size the motor. From a generation standpoint, it is necessary in order to size the generation equipment. For example, if you were constructing a new facility, the electrical utility would need to know the amount of energy you anticipated using, and also the rate at which you would use it. A transformer might supply your energy needs adequately, but if undersized would not meet your power needs.

So your power requirement is the demand you place on the utility, and is identified as **demand** on your utility bill. Power is measured in watts and is usually expressed in kilowatts, which is one thousand watts.

Energy is defined as **the amount of power expended over a period of time**. In electrical usage, it is obtained by multiplying power by time, and is generally expressed in kilowatt-hours. By using conversion factors, it can be converted to an equivalent amount of Btu's, tons of coal, gallons of fuel oil, or other forms of energy.

METERING

Figure 14.1 shows a common type of electrical meter that measures both energy and demand. The pointer measures demand -or power -and is constructed as a ratchet mechanism. It drives upscale as load is applied, and is reset only by the meter reader. If for some reason it doesn't get reset, the highest peak of the combined billing periods is what you will be billed for.

You may have a need to know the instantaneous rate at which power is being consumed. You may wish to know when the peaks are occurring, or how much power a specific piece of equipment needs. This can be done easily with the meter and a stop watch as follows:

- Find the meter constant k on the face of the meter
- Measure in seconds the time of one revolution of the disk.
- Use the equation Power (kw) = k * 3.6/t

Figure 14.1

Demand register incorporated in integrating kilowatt meter. Pointer drives to highest demand for billing period. Is returned manually to zero after reading.

Electrical energy is measured by a kilowatt-hour meter which multiplies the voltage by the component of the in-phase current and then integrates it with respect to time.

The dials on the meter of Figure 14.1 record the energy usage in kilowatt-hrs. Figures 14.2 and 14.3 show how to read these dials. They are not reset at the end of each billing period, so to measure power for a period of time, you must know the beginning readings.

Figure 14.2

Figure 14.3

UNDERSTANDING AND REDUCING ELECTRICAL COSTS 161

In larger industrial and commercial facilities, the demand meter may be of the type that provides a chart indicating magnitude and time of all demands through the billing period. A sampling is taken on either a 15-, 30-, or 60-minute interval, depending on the utility. 15- and 30-minutes are the most common.

Figure 14.4 shows such a chart. It is a plot of power and energy used by an industrial firm for a period of one month. The energy used is represented by the total shaded area. This would show up on the electrical bill as kilowatt-hours (kwh).

The demand (kw) for the month would have been recorded on the 26th of the month. It should be noted that if this facility had been closed every day, but had turned on all equipment for one time interval, the demand meter would have picked it up, and the facility would have been billed for it.

In many cases, the cost for demand is equal to or more than the energy costs, so if this had happened, an electrical bill for the month could still be about half of the usual amount.

Another thing that sometimes happens is a large piece of equipment that is to be tested will be brought on line at peak power time. Sometimes new peaks are reached, and because of ratchet clauses, a penalty is paid for months.

Controlling loads to eliminate peaks resulting in demand penalties is known as **demand control** or **load shedding**. This is done by scheduling - either manually or automatically - large, intermittent loads to operate only at times when other loading is reduced.

In an industrial environment, load shedding is primarily limited to HVAC systems and perhaps air compressors. There may, however, be energy intensive processes which can be rescheduled so they do not coincide with other large users.

Figure 14.4

FROM THE UTILITY STANDPOINT

In order to better understand the rate structure, it is necessary to look at the requirements imposed upon utilities for meeting the electrical needs of the area they serve. In order to obtain the monopoly for electrical service within a given territory, the utility is required to be able to meet the electrical needs of that area at all times.

These needs are constantly varying and are dependant upon many factors. Figure 14.5 shows a typical weekly load curve of a utility. This load is made up of all electrical service provided by the utility and includes residential as well as industrial and commercial. Note that the load fluctuates within each 24-hour period with a high occurring generally somewhere around 2 o'clock in the afternoon and a low sometime around 3 o'clock in the morning. Notice also that the load drops off on the weekend.

The load is heavily dependent upon weather conditions both on a daily and on a seasonal basis. Some utilities have a peak load in the summer when the air conditioning load is highest, while others peak in the winter with heating loads. Unusual weather conditions are usually the reason for a utility reaching a new high in power generation. Figure 14.6 shows both summer and winter peaking curves for one utility. Notice the double peak in the winter curve which is due to heating loads.

In order to design the equipment to handle such loads, the utility will develop a curve similar to that shown in Figure 14.7. This curve is generated by plotting hours versus capacity requirements on a monthly basis. From this information, the designers can break the load into three types - base load, intermediate load, and peak load.

Since the base load is rather constant and makes up the bulk of the total power requirement, the utility will construct a power

Figure 14.5

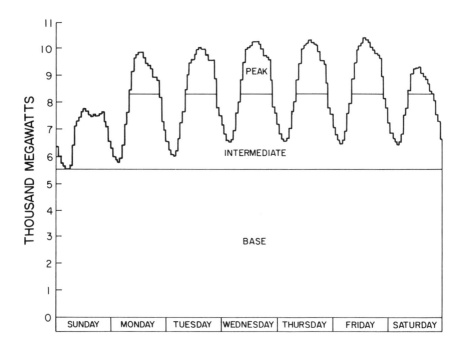

Figure 14.6

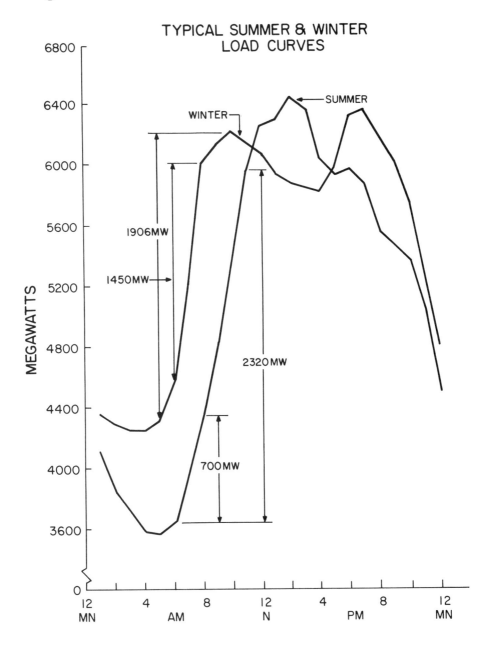

Figure 14.7

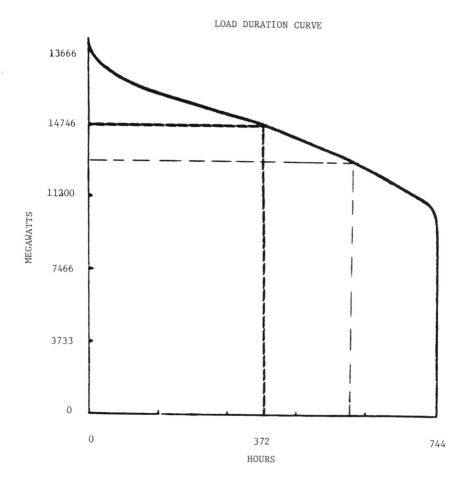

UNDERSTANDING AND REDUCING ELECTRICAL COSTS 167

plant having a high capital investment cost but low operating cost. This will usually be a coal-fired or nuclear powered generating plant.

The intermediate power requirements are met with less capital intensive equipment such as older plants, oil or gas fired boilers, and in some cases hydroelectric power.

Peaking equipment must have the capability of quick start up with relatively short operating time, and may consist of gas-fired turbines or hydroelectric.

Rate schedules then are based upon a return on capital investment and operating expenses - plus some allowable profit.

REACTIVE DEMAND

This component of the rate structure is more difficult to understand than the others, also the least expensive. But a comprehensive knowledge of reactive demand will also help in understanding the factors that influence the efficiency of various pieces of electrical equipment, particularly motors.

In order to understand it, we must first understand phase relationships between current and voltage under different conditions.

<u>Resistive load</u> - First, let's look at the phase relationships of power, voltage and current when applied to a resistive load such as an electric heater. Figure 14.8 shows the relationship that exists between the three. Voltage and current are both in phase with each other. Power is the product of the instantaneous values of voltage and current. Notice that power is always positive. This is so because the voltage and current both swing in a negative direction at the same time, and the product of two negative values is positive.

Figure 14.8

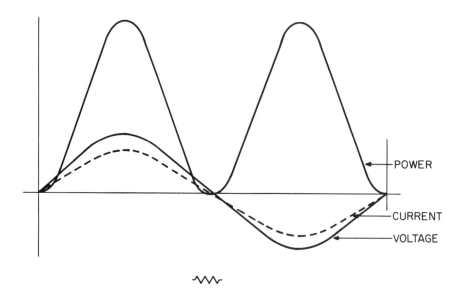

POWER EXPENDED IN A RESISTANCE

UNDERSTANDING AND REDUCING ELECTRICAL COSTS 169

Figure 14.9

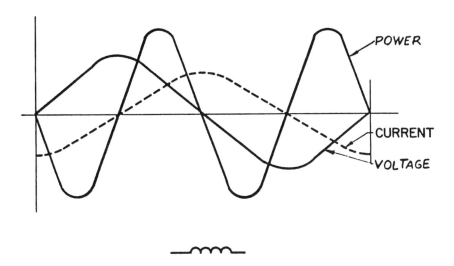

POWER EXPENDED IN AN INDUCTANCE

Inductive load - The phase relationship of voltage, current, and power in an inductive load is quite different from that of a resistive load. An inductive load is created by such components as transformers, motors, or anything having a coil.

When current starts to flow through a coil, a voltage opposite in polarity to that being applied is induced in the coil. This **back electromotive force** as it is called, delays or "chokes" the current passing through the coil. As a result the voltage is caused to lag the applied voltage. If the coil were a pure inductor (a theoretical one) with no resistance, this lag would be 900. This is shown in Figure 14.9.

If we then determine the power expended across the inductor by multiplying the instantaneous voltage and current, we see that the power cycles go both negative and positive and are of equal magnitude. The result is that the net power expended across a theoretical inductor is zero. What happens is the inductor absorbs energy from the power source on one half cycle and establishes an electromagnetic field. This field then collapses around the inductor on the next half cycle, and the energy is returned back into the circuit. The net result is a phase shift between current and voltage, but no power is expended.

In reality, it is impossible to have a pure inductance without some resistance. Therefore we are looking at a combination of resistance and inductance when we consider electrical components that have coils, such as electric motors.

Figure 14.10 shows the phase relationships involving both resistance and inductance. The current is no longer 900 out of phase with the voltage, but is at some less value.

This out of phase power performs no useful work but does increase the total amount of current required to give an equivalent amount of work.

UNDERSTANDING AND REDUCING ELECTRICAL COSTS 171

<u>Capacitive load</u> - If we hook a capacitor across a power circuit and again look at the phase relationship between voltage, current, and power, we discover that again they are 900 out of phase, but in the opposite direction to that experienced in an inductor. This is shown in figure 14.11. This phase shift is explained by the fact that the current must flow and build up in the plates of the capacitor before the voltage builds up.

<u>Power factor</u> - Because of the great number of electric motors, power companies are concerned about the phase shift caused by the inductive load. Therefore, in some rate schedules, there is a charge for lagging reactive demand.

If we plot the phase relationship between real power and reactive power vectorially, we can determine the phase angle and a quantity known as power factor. This is shown in Figure 14.12 along with equations necessary to calculate power factor. The ideal situation is to have the power factor approach a value of one. In reality most power companies do not impose a reactive demand charge if this value is as high as .90.

The procedure to calculate power factor from an electric bill is a follows:

> The two things that are known or are available from the power bill are kilowatts of demand and kilovars of reactive demand. Our object is to determine the cosine of the phase angle 0 (Fig. 14.12) which is the power factor. But first, from the two values known we can only determine the tangent of the angle.

Tangent 0 = kvar/kw

After determining the tangent of 0, the angle 0 is then determined from trigonometric tables or a calculator that has trigonometric functions. From this then the cosine is determined, which is the power factor. An example is given below.

Figure 14.10

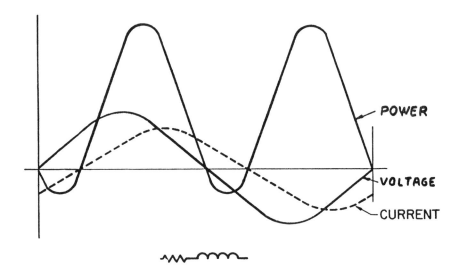

POWER EXPENDED IN RESISTOR AND
INDUCTANCE (ELECTRIC MOTOR)

Figure 14.11

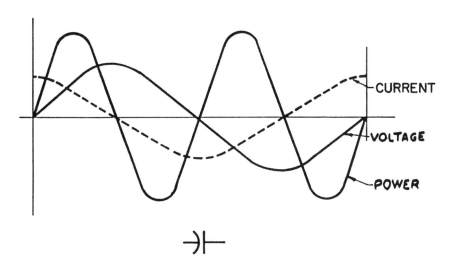

POWER EXPENDED IN A CAPACITANCE

Figure 14.12

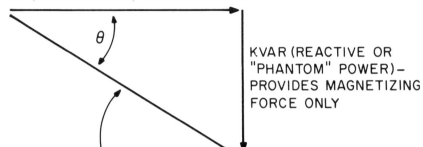

θ = PHASE ANGLE = MEASURE OF NET AMOUNT OF INDUCTIVE REACTANCE IN CIRCUIT

$$\cos\theta = \frac{KW}{KVA} = P.F. = \text{RATIO OF "REAL POWER" TO "APPARENT POWER"}$$

$$\sin\theta = \frac{KVAR}{KVA}$$

$$\tan\theta = \frac{KVAR}{KW}$$

$$KVA = \frac{KW}{\cos\theta} = \frac{KW}{P.F.} = \frac{KVAR}{\sin\theta} = \sqrt{(KW)^2 + (KVAR)^2}$$

$$KW = KVA\cos\theta = KVA\,(P.F.) = \frac{KVAR}{\tan\theta} = \sqrt{(KVA)^2 - (KVAR)^2}$$

$$KVAR = KVA\sin\theta = KW\tan\theta = \sqrt{(KVA)^2 - (KW)^2}$$

UNDERSTANDING AND REDUCING ELECTRICAL COSTS

From an electric bill we read kw demand equals 1440. Kvar demand is 1230.

Tangent 0 = kvar/kw = 1230/1440 = .854

Arc tan of .854 = 400

Cosine 400 = .76

Therefore: power factor = .76

<u>Penalties for poor power factor</u> - Poor power factor can penalize the user in more ways than just a slight increase in the electric bill. For example:

It unnecessarily increases the load on the electrical distribution system. This can cause overloads, can limit the capacity for expansion, and creates what is known as I2R losses. These are losses due to the resistance in the wire, which admittedly are usually small, but notice that they are proportional to the square of the current.

The higher current can cause excessive voltage drops in the line, so power to the end using component may be inadequate.

In some cases utilities have a heavy penalty for a low power factor. This seems to be more true of co-ops.

<u>Correction for lagging reactive demand</u> - There are several ways to improve the power factor. Some of these are listed below:

1. Capacitors are by far the simplest and most versatile scheme for power factor improvement. As determined previously, the phase relationship of voltage and current in a capacitor is opposite to that in an inductor, Therefore by matching the capacitive load to the inductive load, the overall reactive effect can be cancelled. It is best to locate capacitors right at the inductive load. However, from a practical standpoint, satisfactory results can usually be obtained by overall capacity correction at the power input to the facility.

2. Synchronous motors can be run with a leading power factor which will cancel inductive effects. However this is practical only in very large installations where several hundred horsepower and up are required. In addition, they must be run continuously to correct the power factor.

3. Fluorescent lights have a capacitive load. Some energy managers have discovered this when they began to cut back on lighting only to find the power factor decreasing.

4. There is equipment on the market that continuously senses the power factor and automatically increases or decreases the capacitive load to keep the power factor at a constant high level.

5. The manner in which electrical equipment is utilized can affect the power factor. For example, induction motors at full load have better power factors that those running at part or no load. The reason for this is that the reactive component (kvar) which provides the magnetizing force remains virtually constant from no load to full load, while the demand (kw) component is a function of loading.

ANALYZING YOUR ELECTRIC RATE STRUCTURE

It is of the utmost importance that you have a good understanding of the rate schedules of your electrical utility, and any supplemental contracts, before you begin an electrical energy conservation effort. It is entirely possible that without this knowledge you could do things to save energy without impacting on the electrical bill. You should also realize that you may take steps that will save money, but will not necessarily save energy. Reducing demand - or load leveling - is an example. Changing rate schedules is another.

Some utilities will have a separate contract that specifies the size of equipment to be supplied and other details. The contract may also specify a lower limit for demand, somewhat like a ratchet clause.

Your strategy for reducing electrical costs should be developed around the rate structure and any contracts associated with it.

Figure 14.13 is an example of this. It is an hourly plot of demand and has a peak of 290 kw. If you were on a ratchet clause that required you to pay for 90% of the demand reached during a certain period, then anything you do to reduce demand below that level will have no payback until the next cycle. If your contract states that your minimum billing for demand will be 60% of that contracted for, then dropping below 160 kw will have no payback. However, you may be able to re-negotiate your contract.

You should have access to all the rate schedules of the utility and become familiar with them for two basic reasons. You need first of all to determine if you are on the best schedule for your operation. This is a customer election, and the original basis for selection may have changed - such as could happen with growth, or simply having two or more meters combined into one.

Secondly, you need to know what additional options may be available to you. For example, you may have an operation with a heavy electrical load that you would be willing to put on an interruptible service - if the utility would make it worth your while. Or you may be considering replacing a steam reducing valve with a turbine and generator, which could classify you as a cogenerator. The energy manager of a large chemical company did this, and basically found a loophole in the rate structure that allowed him to have a two months payback on the whole cogeneration project.

HOW TO ANALYZE YOUR ELECTRICAL ENERGY

It is possible just from data on the electrical bills to do a detailed analysis of the electrical energy use in a facility without even surveying the facility, or having much knowledge of its operation. Such an analysis should be the first step in conducting an electrical audit. And of course, the best way to analyze the data is to display it graphically.

178 MANAGING ENERGY RESOURCES IN TIMES OF DYNAMIC CHANGE

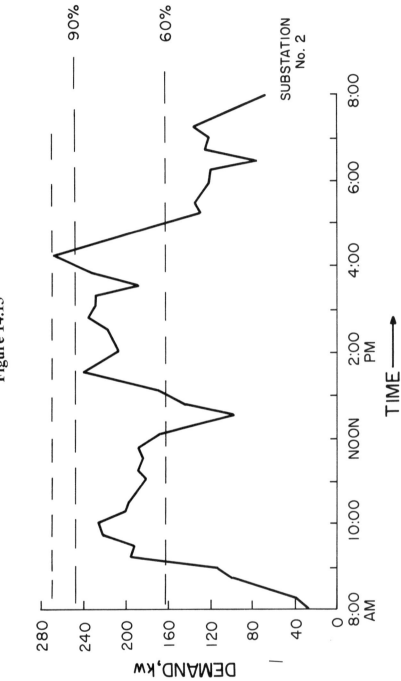

Figure 14.13

UNDERSTANDING AND REDUCING ELECTRICAL COSTS

From the data on the electrical bill, you can graph the following information:
- Energy use
- Demand
- Load factor
- Utilization factor
- Power factor

Figures 14.14 through 14.18 are graphs of the above information taken from the electric bills of a medium sized wood products company. Let's look at how much information we can amass just from this data.

From a plot of the electrical energy used (kwh) in Figure 14.14, we see no seasonal fluctuation, so it is obvious there is no significant air conditioning or electrical heating loads. We can quickly discount these as a potential for improved efficiency, better insulation, or load shedding.

The steady increase in energy use with time indicates growth of the company with an increasing production, which, as it turns out, was the case. The company had recently moved into the area and was experiencing a growth in sales and production.

The demand curve of Figure 14.15 indicates new equipment being added to meet increased production demands. By examining the power factor curve of figure 14.18, we see that when the demand went up, the power factor went down slightly. This indicates an inductive load, most probably large motors. The question then arises - should a second shift be added rather than buy additional equipment which will drive up demand.

The load factor curve of Figure 14.16 and the electrical utilization curve of Figure 14.17 provide answers to this. <u>Load factor</u> is

Figure 14.14

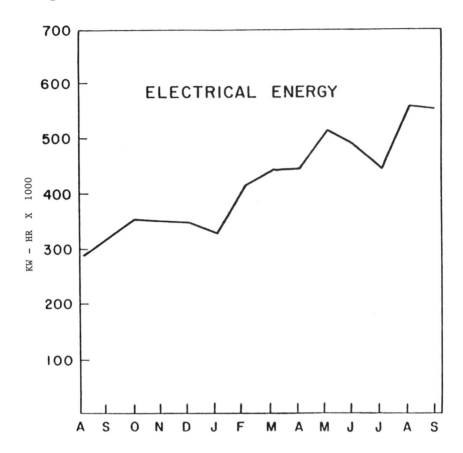

Figure 14.15

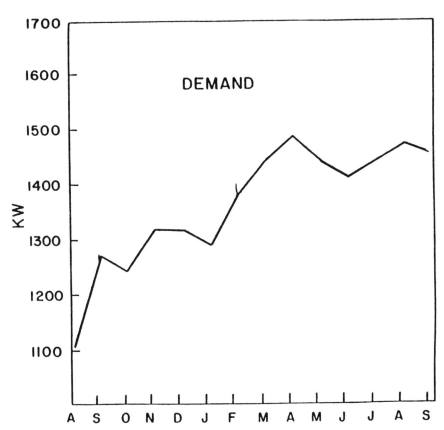

an indication of how much of the total energy available is being used. It is obtained by dividing the total energy used per month (kwh) by that amount that would have been used if it had been consumed at maximum demand for every hour in the month.

While it is good to have a high load factor, the numerical values are not as important as the constant values. If there are wild swings in this number from month to month, it indicates high demand or low utilization of the equipment - or both. For example, if a plant closes down for vacation for a week or two during the month, the load factor will dip to reflect this.

Another load factor curve can be developed whereby the demand is multiplied by the <u>actual</u> operating hours during the month. This value then is divided into the total kilowatt hours used per month to determine the load factor. This number should be slightly under one since you probably do not operate at the peak demand all the time. If it is over one, this indicates that electrical equipment is running beyond working hours.

Notice in this case it is fairly constant, and is in fact improving as production increases. This indicates that a second shift is being utilized, at least for part of the production.

This is confirmed in Figure 14.16, the <u>electrical utilization</u> graph. The points on this curve are obtained by simply dividing kw of peak demand for the month into kwh of energy for the month. The answer is in hours, and represents the number of hours you would have to run the equipment at peak demand in order to use the same amount of energy you actually used. In an average working month with a one shift operation there are 172 working hours. The first dashed line in Figure 14.16 represents then a single shift operation. Two shifts are represented by the second dashed line at 342 hours.

Notice that the electrical utilization is somewhere between the first and second shift, with one peak going into the second shift.

This was discussed with the plant management, and it was determined that there was some second shift operation, but not as much as indicated by the graph. Then we started to look for a culprit. A substantial amount of equipment was being operated after normal working hours.

After some exploration, it was determined that maintenance people were leaving large blower fans running after hours to make sure all the explosive dust was removed from the plant. And they were coming in very early to start up the equipment long before production started. Maintenance work was also done on the weekend in which the equipment was run for a considerable length of time.

After seeing this information displayed graphically, management was able to reduce the overall running time of large pieces of equipment by re-scheduling and simply making people aware of the cost of operation of the equipment.

All of this was achieved without really having to get intimately involved in the operation of the company, but by simply flagging potential savings and asking the right questions.

Another interesting example involved a sawmill operation. Their electric bills were sent to me for an analysis and some suggestions on methods of reducing costs. The company had three separate meters for which they were being billed. I had no knowledge of their location, whether they were close together in the next county. But, I suggested their first consideration should be to combine them into one if at all possible. The total cost for demand would go down if the peaks weren't appearing simultaneously. Also, the declining step rate for kwh would provide a savings. It was also possible that they would be eligible for a new rate schedule which could be beneficial.

Figure 14.16

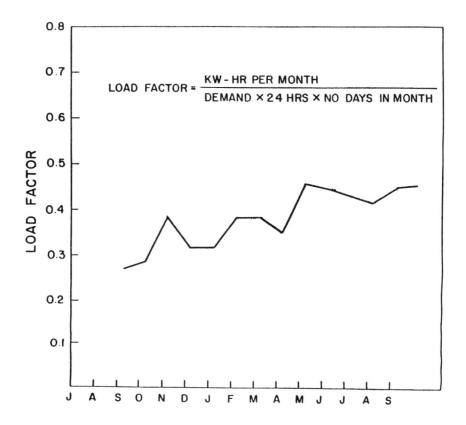

Figure 14.17

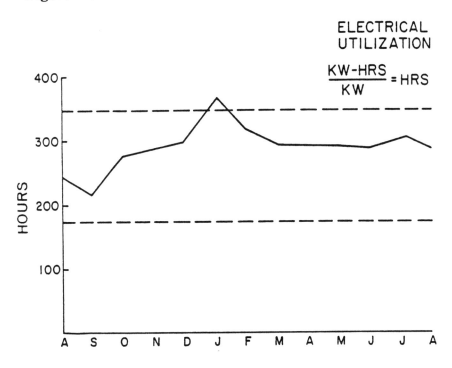

Figure 14.18

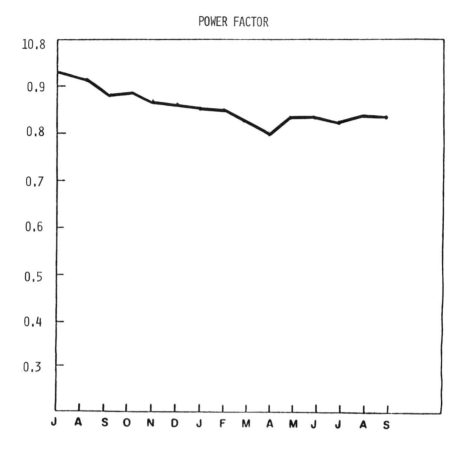

By dividing kwh by kw, I determined the electrical utilization factor in hours for each meter as follows:

M_1 75,000 kwh/550 kw = 136 hours
M_2 118,000kwh/675 kw = 174 hours
M_3 3,200kwh/150 kw = 21 hours

The operation at the first two meters is probably a one shift operation. However, that at meter three is obviously operating part time during the month. The total bill for the third meter was $1403. Dividing this by the 21 hours gave $70 per hour for electrical costs for this operation.

It is obvious that the operation at meter 3 should be combined with the other two for billing purposes, or else the value of the product produced there evaluated very carefully.

These are examples of savings located just from analyzing the electric bill. If your utility can provide you with a 15 or 30 minute breakdown of demand and energy use -or if you can measure it yourself - you may be able to do further analysis such as pinpoint pieces of equipment that are used for a very short duration at peak times.

CONDUCTING AN ELECTRICAL SURVEY

There are computer programs that can provide much assistance in managing energy, such as those that will analyze the energy use of a building, or size equipment, etc., but none for conducting an electrical survey because it is simply a matter of counting and measuring. A survey is not an audit in which energy conservation opportunities (ECO's) are developed. It is simply an accounting of the electrical energy.

There are three prime sources, or methods for developing this information. They are: your electric utility, electrical measurement instruments, and nameplate information.

THE ELECTRIC COMPANY - The place to start an electrical survey is at the point it enters your facility. You may already have enough metering there to give you all the information you need. You can obtain from your meters the total energy for any period of time, instantaneous demand (by timing the disk cycle), peak demand, and power factor. This is assuming you are on a rate schedule with these charges.

Many times, however you wish to know not only the total power but also that being consumed in various sublevels, or legs. These may be heavy currents and high voltages that would require not only expensive meters, but experience in using them. This is where your electric utility may be of help. They do this sort of thing routinely and hence have the equipment and experience. There is usually a charge for this kind of service, but it is usually nominal.

In fact, they may already have much of the data you wish. Sometimes this information is necessary for their planning. So before you start any kind of measurement program, find out how much they already have. This is particularly true if you are looking at demand. The utility probably has a magnetic tape of your demand by intervals and can supply it to you. With most utilities, anything you are doing to reduce demand will get their full support.

INSTRUMENTS - If you have found a good way to justify the purchase of instruments for measuring energy, the rest of us would like to know. My only suggestion has been to have a good energy policy, then lay claim to the necessity for instruments in order to meet the requirements of the policy. Anyway, it is difficult if not totally out of the question, so here are some alternative ideas.

If there are sister - or brother - organizations that have the same need, perhaps they could be purchased at a central level for use by everyone.

There are several firms that have instruments for rent. General Electric is one that comes to mind. I am sure there are others.

If there is a university nearby, and you have contributed regularly to the college of engineering - rather than the athletic department - you can sometimes borrow some of their equipment.

There are several brands of instruments that you can simply hook across the lines and collect this information. One precaution - be sure if you want power measurements you are not getting just volts times amps. The power factor must be taken into account, and there is equipment that professes to measure power, but in truth measures only volts times amps.

<u>WHAT KIND OF INFORMATION</u> - In compiling the information for an electrical survey, you will need the following information:

- The horsepower of each motor
- The total lighting load
- A list of all discretionary loads
- An operations schedule for each of the above.

Motor horsepower is most accurately obtained with a power meter. But even then, the motor may be operating with a varying load, so some judgement must be exercised to come up with both the peak and an average. Another way is to simply take the horsepower reading from the nameplate. Again some judgmental factors must be applied, because very few motors operate at their rated load. On an average they will operate at 70% of rated load.

If you have a better figure for your operation, use it. Then there is the matter of efficiency which again you must assign a value to. A rule of thumb that makes the whole process easier and one you may wish to use is as follows:

One horsepower = .746 KW

74.6% is a little low for the efficiency of an electric motor, but if we use it, then:

One horsepower = One Kilowatt of power

Here is an example of using the above rules of thumb to calculate actual electrical power from nameplate information.

Nameplate horsepower = 400

Assume motor operates at 70% load

.70 * 400 = 280 HP

1Hp = approximately 1KW

Therefore: 280 HP = 280 KW

The lighting load is more straight forward. Simply read the power rating from the bulb, count the bulbs, and total up.

It is important to seek out and list discretionary loads because these are the ones that can possibly be shed in times of high demand. These are loads that can be interrupted without a detrimental effect, or loads that have a flywheel effect such as HVAC, refrigeration, etc. You will probably have more discretionary loads than you think. The first reaction of most people is that they have an insignificant amount. Some examples of discretionary loads are:

- fans
- HVAC
- refrigeration
- hot water heaters
- air compressors
- lights

The operational schedule of each piece of equipment should be recorded. By plotting operating time vs power for each, then imposing them on each other you may, by simply changing operating times - perhaps with a timer - significantly reduce your demand.

CONCLUSION

Electrical costs are at present rising faster than other energy sources. Some are taking quantum jumps as new generating facilities come on line. This means that a good understanding of your electrical charges and ways in which they can be reduced is of prime importance.

Appendix A

SAMPLE ENERGY POLICY

BRAND X MANUFACTURING COMPANY

POLICY AND PROCEDURES MANUAL

SUBJECT: ENERGY MANAGEMENT PROGRAM

I. POLICY

Energy management shall be practiced in all areas of the Company's operations.

II. ENERGY MANAGEMENT PROGRAM OBJECTIVES

It is the Company's objective to provide energy security for the organization for both immediate and long range situations by:

- Utilizing energy efficiently
- Incorporating energy efficiency into new and existing equipment and facilities.
- Complying with government regulations - Federal, state, and local.
- Putting in place an Energy Management Program to accomplish the above objectives.

III. IMPLEMENTATION

A. Organization

The Company's Energy Management Program will be administered through the Planning Department.

1. Energy Manager

The Energy Manager will report directly to the Director of Planning, and shall have overall responsibility for carrying out the Energy Management Program.

2. Energy Committee

The Energy Manager may appoint an Energy Committee to be comprised of representatives from various departments who will serve for specified periods of time. The purpose of the Energy Committee is to advise the Energy Manager on the operation of the Energy Management Program.

3. Energy Coordinators

Energy Coordinators shall be appointed to represent a specific department or division. The Energy Manager shall establish minimum qualification standards for Coordinators and shall have joint approval authority for each Coordinator appointed.

They shall be responsible for maintaining an ongoing awareness of sources of energy consumption and expenditures in their assigned areas. They shall recommend and implement energy conservation projects and energy management practices.

Coordinators shall provide the necessary information for reporting from their specific areas.

They may be assigned on a full-time or part-time basis as required to implement this program in the specific area.

B. Reporting

The energy Coordinator shall report all efforts to improve energy efficiency to the Energy Office as these efforts are undertaken. A summary of energy cost savings actions for each quarter shall be submitted by the Coordinator to the Energy Office.

The Energy Office will be responsible for consolidating reports as required by management.

C. Training

The Energy Manager shall provide energy training at all levels of the Company.

IV. POLICY UPDATING

The Energy Manager and the Energy Advisory Committee shall review this policy annually and make recommendations for updating or changes.

Appendix B

THE ENERGY MANAGEMENT PROGRAM AT SOUTHWIRE

John T. Kopfle, P.E., C.E.M.
Senior Energy Engineer
Corporate Energy Management Department
Southwire Company
P.O. Box 1000
Carrollton, GA 30119-0001

INTRODUCTION

Southwire Company, Carrollton, GA, manufactures copper and aluminum wire, cable and rod for commercial and residential buildings, utilities, electronic manufacturers and others. Southwire is the nation's largest privately-owned rod, wire and cable manufacturer and operates 12 plants in six states and employs over 3,500 people. The plant locations and products are given in Table I. In addition to its wire and cable plants, Southwire owns an electrolytic copper refinery, a machinery division, an automated sawmill and a hydroelectric plant.

Southwire's Corporate Energy Management (CEM) department has the task of reducing the energy component of manufacturing costs and assuring adequate supplies of energy for all the facilities mentioned above.

ORIGIN OF ENERGY MANAGEMENT PROGRAM AT SOUTHWIRE

With the quadrupling of oil prices as a result of the 1973 Arab oil embargo, the economics of manufacturing changed drastically for many firms. Companies began serious efforts to reduce energy consumption and remain competitive. Southwire joined

TABLE I

SOUTHWIRE MANUFACTURING FACILITIES

FACILITY	LOCATION	PRODUCT
Carrollton Utility Product Plant	Carrollton, Ga.	Utility Wire & Cable
Carrollton Building Wire Plant	Carrollton, Ga.	Building Wire
Copper Division Southwire	Carrollton, Ga.	Cathode Copper
Copper Rod Division	Carrollton, Ga.	Copper Rod
Southwire Machinery Division	Carrollton, Ga.	Machine Shop Services
Wood Products Division	Carrollton, Ga.	Lumber, Pallets
Kentucky Rod & Cable Mill	Hawesville, Ky.	Aluminum Rod & Cable
Arkansas Building Wire Plant	Osceola, Ark.	Building Wire
Southwire Specialty Prod. Div.	Osceola, Ark.	Specialty Wire
Flora Utility Products Plant	Flora, Ill.	Utility Wire & Cable
Wyre Wynd Division	Jewett City, Conn.	Specialty Copper Wire
Utah Building Wire Plant	West Jordan, Utah	Building Wire

APPENDIX B 199

FIGURE 1

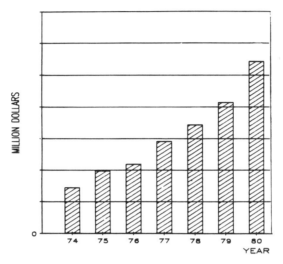

FIGURE 2

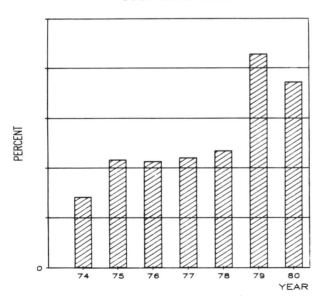

in this effort by forming an Energy Task Force to begin audits, metering, and the application of engineering talent to the problems of energy use. The effort was strongly supported by company founder Roy Richards, himself an engineer, who realized the importance of minimizing increases in manufacturing cost.

The Energy Task Force continued until 1980, by which time the second world oil price shock had occurred. Southwire became concerned by its rapidly increasing energy costs (Figure 1) and, more importantly, by the large increase in the relationship of energy cost to conversion cost (Figure 2). Although it constitutes a small percentage of cost of sales, energy is a large component of conversion cost. Conversion cost which includes controllable costs such as energy, labor and maintenance is more significant than cost of sales. The major portion of cost of sales is material cost, which is largely uncontrollable. Southwire saw that if the trend in energy cost continued, profitability would be seriously eroded.

Due to these concerns, in 1980 the company formed the CEM department, consisting of the corporate energy manager, energy engineers and technicians. During the years of its existence, the CEM department has played a large part in holding down Southwire's cost of energy.

ORGANIZATION OF CEM DEPARTMENT

Figure 3 shows the organization of the CEM department. The department is divided into two functional groups: operations and projects. This division allows the department to devote relatively equal time to managing existing energy projects and developing new projects.

The operations group is primarily concerned with overseeing Southwire's energy management system (EMS) and hydroelectric plant. EMS terminals have now been installed in all of Southwire's plants nationwide, and are monitored and controlled in Carrollton via use of modems. Load shedding capability through use of the EMS is being constantly updated to allow Southwire to obtain maximum savings on electrical demand charges while minimizing disruptions to plant operations.

The hydroelectric plant, which is located adjacent to Southwire's Connecticut manufacturing facility, has operated for almost three years. The power produced by the plant displaces power purchased from the utility and lowers the manufacturing facility's cost of electricity. The EMS allows the operation of the hydroelectric plant to be monitored from our Carrollton office. The operations group also supervises maintenance and improvements to the plant.

The projects group works in developing and implementing energy projects. The development task includes searching out areas of high energy usage; determining ways to reduce that usage; specifying the equipment required; and calculating the economics of the proposed change. After a project has been approved, the project implementation function begins. This involves overseeing the specification, purchasing, installation and startup of the equipment.

There are other Southwire personnel who operate under the supervision of the CEM department, as shown in Figure 3. These include the EMS coordinators at each plant, the operators of the hydroelectric plant and the operators of the peak shaving engines. These personnel spend just part of their time doing energy-related work, and are not formally part of the CEM department.

Each of the outside plants typically has one person who acts as energy coordinator. Although this is not a formal distinction, nevertheless these people provide a valuable service in data gathering, assistance in project implementation and in general facilitating the work of the CEM department.

FIGURE 3

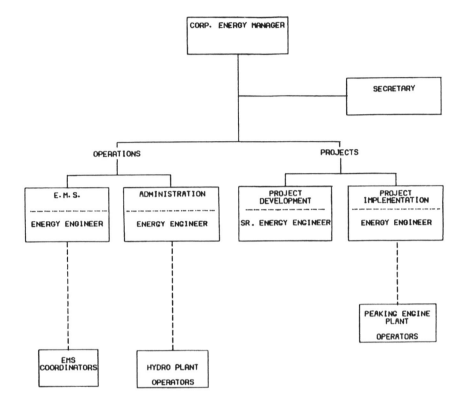

TYPICAL PROJECTS

During the seven years of its existence, the CEM department has been involved in the whole gamut of energy management projects, including energy audits, lights, air conditioning and heating, motors, electrical demand control, utility rate negotiation, power generation, fuel storage, burner and furnace retrofits, cooling towers, boilers, and many others. A comprehensive list of ongoing efforts is given in the Action Plans section of the paper, "Southwire's Stategic Energy Plan." Probably the three most comprehensive and interesting projects to date are the energy management system (EMS), the hydroelectric plant and the peak shaving engines.

Energy Management System - The EMS was installed in 1982 for monitoring electrical loads and reducing electrical demand. The system uses distributed intelligence technology, meaning that monitoring of electrical loads, decision-making and controlling are done by multiple computers which can interact with each other. As Figure 4 shows, the heart of the system is the central intelligence units (CIU's), which perform the decision-making and control functions.

Each CIU can control as many as 16 imput/output units, which are connected to the loads to be controlled. The CIU's are programmed from a terminal in Southwire's Carrollton office.

Since the EMS was installed, we have continually refined our load control procedures. When it is necessary to control our load (typically during hot summer afternoons), the EMS begins shutting down equipment until the load setpoint is reached. Close coordination with plant operating personnel is required to determine the order in which equipment is turned off. After working with them several years, we have good cooperation from plant operators, especially when they are made aware of the savings possible.

FIGURE 4

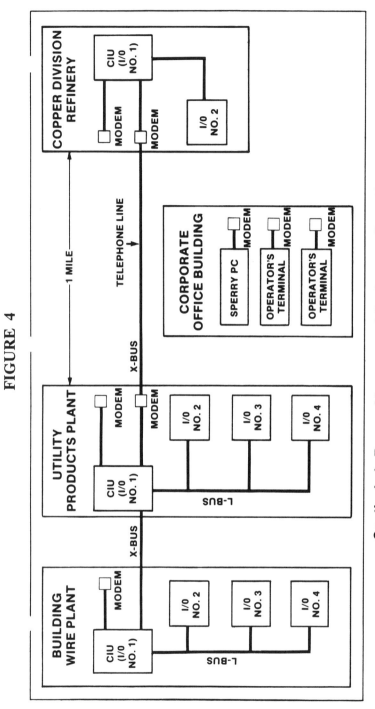

Southwire's Energy Management System configuration

A personal computer is used in conjunction with the EMS to gather data automatically. The computer also manipulates the data for evaluation purposes.

The EMS units located at Southwire's remote plants are accessed by modem from the Carrollton office. These units are used for data gathering purposes and load control. As with the Carrollton Plants, load control capabilities are being constantly refined.

The benefits of load control via the EMS can be substantial. Many utilities have electric rates which contain "ratchet" provisions, meaning that a peak demand set one time is paid for all year long. Use of an EMS for load shedding can reduce the peak demand and provide year-round savings. For instance, given a typical demand charge of $100 per kilowatt per year, shedding one magawatt of load can provide savings of $100,000 per year. Since EMS equipment is relatively inexpensive, excellent paybacks can result.

Hydroelectric Plant - Southwire's Connecticut manufacturing facility is located in a converted textile mill, which was built in the late 1800's and used water from the adjacent river to power its machinery. In the early 1900's, a turbine was installed to generate power for the textile mill. This turbine operated until 1969, when the electrical equipment was destroyed by fire. The facility was not rebuilt due to the low power rates at the time.

By the early 1980's power rates in Connecticut had increased substantially and a feasibility study was performed on rebuilding the facility and again generating power. The study showed that, given the large expected increases in power rates due to fossil fuel increases and nuclear plant capital costs, rebuilding of the hydro facility was justified. Construction on the project began in 1983, and included a complete rebuild of the powerhouse, repair of the existing civil structures and installation of a state-of-the-art 2.6

megawatt turbine/generator. The facility was started up in February 1984 and has operated reliably since that time.

The turbine/generator is controlled using a programmable controller, which is programmed to optimize output given the characteristics of the river flow and turbine. Water level, electrical output and other factors are monitored from the Carrollton office using the EMS. Personnel on site oversee the daily operation, troubleshoot and perform routine maintenance.

The generator is tied into both the adjacent manufacturing plant and the utility grid. All power produced is used at the manufacturing plant unless the generator output exceeds the plant load, at which time the excess power is fed through the plant grid to the utility grid. During those times when plant demand is greater than the hydro output, the additional power is purchased from the utility.

Savings due to the hydro plant have been substantial, as it supplies about two-thirds of the power used at the manufacturing plant. Through careful management of operations and maintenance, it should continue to operate reliably for many years.

Peak Shaving Engines - Although the EMS does a good job of load shedding, there is a limitation on the reduction we can achieve through simply shutting off machines. Many of Southwire's processes are either on or off, with no ability to partially reduce electrical consumption. Also, due to production demands it is not fesible to shut off certain equipment even for a few hours each summer.

Despite these concerns, there was still more potential for demand reduction and its substantial savings. Accordingly, Southwire looked at the possibility of installing engine/generators which would operate during peak periods and allow us to capture savings beyond that possible using the EMS alone. Three natural gas-fired engines with a total capacity of 3.9 megawatts were installed in May 1984 and have operated during the past three summers.

The peaking engines work in conjunction with the EMS. During peak demand periods, the utility sends a signal via modem notifying us of the timing and duration of a curtailment period. Through the combination of the EMS and the peaking engines, we reduce the electrical load the utility sees down to our billing demand level, and maintain it there for the duration of the curtailment period. Although it is a challenge to keep the load down, it is required for only a few hours per summer. As described previously, the savings can be substantial.

Recently the EMS has been configured to control the generator output either manually or automatically. This allows us to run the engines at just the level required to stay below the billing demand.

RESULTS OF SOUTHWIRE ENERGY MANAGEMENT PROGRAM

There are two means of lowering energy cost: conservation and cost avoidance. Conservation is simply reducing the amount of energy used, through such means as: use of high efficiency lights; specifying energy efficient motors; upgrading burners; and submetering to place responsibility for energy use on end users. Cost avoidance refers to reducing the unit cost of energy through: use of EMS and peaking engines to control demand; pursuit of favorable power rates; negotiation with utilities; and self purchase of natural gas.

Both conservation and cost avoidance should be aggressively pursued to reduce energy cost. At Southwire during the early stages of our energy management efforts the primary focus was on conservation, probably because rising energy prices were seen as inevitable and uncontrollable. Given that assumption, reducing the amount of energy used is the only way to reduce total ener-

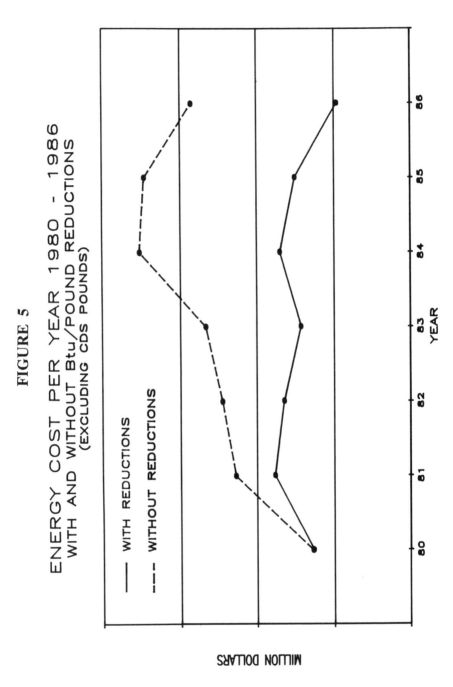

FIGURE 5

ENERGY COST PER YEAR 1980 - 1986
WITH AND WITHOUT Btu/POUND REDUCTIONS
(EXCLUDING CDS POUNDS)

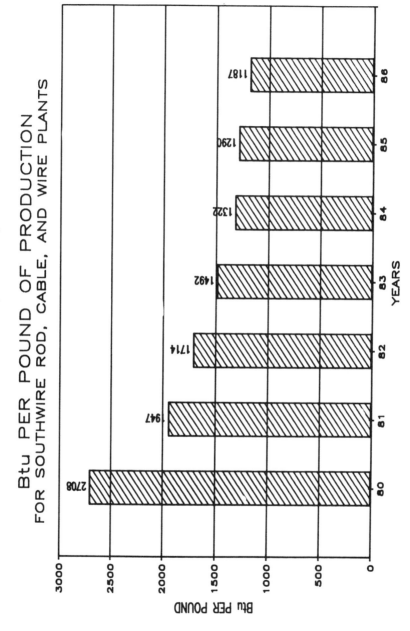

FIGURE 5 (continued)

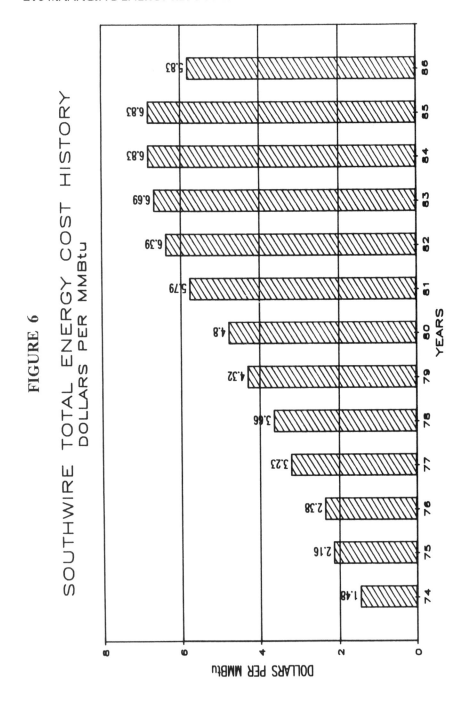

FIGURE 6

SOUTHWIRE TOTAL ENERGY COST HISTORY
DOLLARS PER MMBtu

gy cost. During the last two years, however, more and more of the CEM department's effort has been devoted to cost avoidance measures. One reason is that most of the obvious conservation measures have been done, and it is getting more difficult to further reduce energy use. A second reason is the emergence of cost avoidance opportunities due to innovative power rates, the partial deregulation of the natural gas industry, improved EMS technology and other factors. This same phenomenon has been recognized by energy managers nationwide. There is an increasing realization that cost avoidance measures should be pursued so that no potential savings are "left on the table."

The results of Southwire's conservation efforts from 1981-86 have been significant. Figure 5 shows that company-wide energy usage per pound of product produced (our measure of energy efficiency) was reduced by almost 40% over that period, resulting in a cumulative savings of about $40 million. Our goal is to reduce energy usage by 3% per year for the next two years.

Figure 6 shows that due in part to the cost avoidance measures described previously, the increase in cost per unit of energy has leveled off over the past few years and actually decreased in 1986 to the 1981 level.

The combination of conservation and cost avoidance savings has resulted in substantial dollar savings for Southwire over the past seven years. Figure 7 is a graph of total energy cost for the company. It is significant that the reduction in Southwire's total energy cost from 1981 to 1986 occurred while production nearly doubled. The dollars freed up as a result of the reduction in energy cost over the last two years add directly to profit, which indicates the importance of the energy-saving efforts. Whereas only a small fraction of increased sales dollars goes to the bottom line, all energy cost savings do, and thus emphasis placed on energy cost cutting is generally very worthwhile.

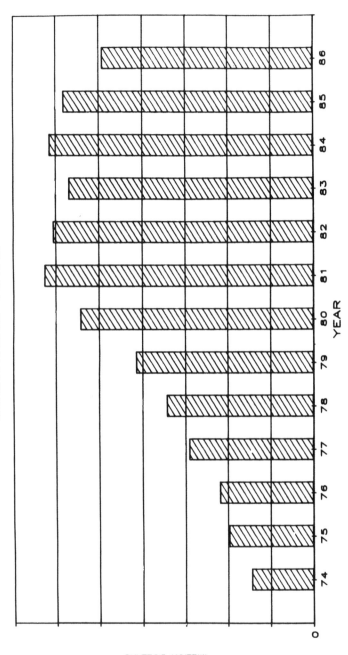

FIGURE 8

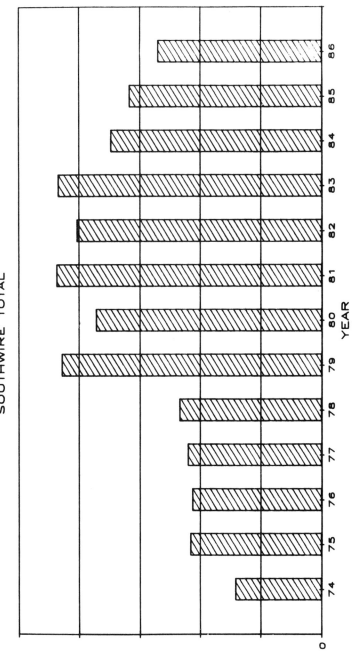

Perhaps the most important relationship is shown in Figure 8, energy cost to conversion cost. As noted earlier, conversion cost is crucial to a firm such as Southwire because it is somewhat controllable and can gain us a competitive advantage. As a result of the energy cost savings realized over the past several years, energy as a percent of conversion cost has reversed its ascent and is now decreasing to the level of the mid-1970's. Firms which are able to decrease the energy component of manufacturing cost have a major advantage over those which do not.

There is one final point worth bearing in mind with regard to energy cost savings. The heart of the mission of an energy management department is saving money, not necessarily energy. This is a point which can be overlooked in the zeal to conserve energy. As the cost avoidance savings figures for Southwire show, both conservation and cost avoidance must be pursued.

STRENGTHS AND WEAKNESSES

Part of developing a Strategic Energy Plan (SEP) for Southwire was the assessing of Southwire's strengths and weaknesses with regard to energy. The first step was to look at the company's energy usage by end use and energy source. Figures 9-12 show Southwire energy consumption broken down in four different ways.

The following comments came from the SEP and give a good indication of areas in which the program is strong and areas in which improvements can be made.

Management Commitment - From the start of Southwire's energy conservation efforts, the commitment of management has been strong. This is an important factor because without this commitment it is difficult to get the cooperation needed for a successful energy program.

A recent article in Energy User News stated that of 19 large industrial companies which had full-time energy managers in 1980, by 1985 only eight of them remained. In addition, "most of those eight expressed a mood of frustration over the creeping energy apathy that was reinfecting their upper management". With that sort of situation occurring, it is impossible to continue the progress of an energy program. At Southwire management commitment has remained strong through the "energy crisis" and current oil glut. This commitment has manifested itself in resources directed toward energy management, operations changed due to energy considerations, concern of line management about energy use, and in other ways.

Accountability - Experience has shown that making operating departments accountable for their own energy use can be the most effective way to reduce overall energy consumption. At Southwire this has been accomplished by accurately submetering each manufacturing unit's energy use and billing them for that use. This helps ensure that the unit takes responsibility for reducing its energy use.

We have improved metering so that in most cases there are sufficient meters and they are accurate enough to provide a good indication of energy use. In some areas, however, such as steam and compressed air, more can be done with regard to submetering and improved accuracy.

The lack of good meter data is an especially insidious problem because it often goes unnoticed. Steam and compressed air each cost Southwire about $1 million per year at Carrollton, but not enough has been done to determine how much should be spent in these areas, because the manufacturing units have been paying an allocated or estimated bill each month.

Another advantage of good metering is that it enables us to accurately bill each plant for the energy they use, as was discussed above. The more accurate the meters, the better correlation

FIGURE 9

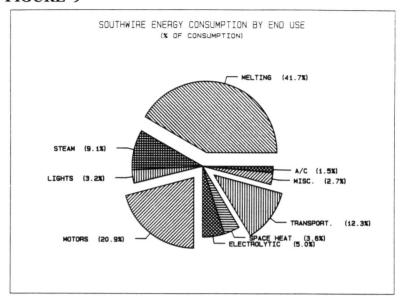

FIGURE 10

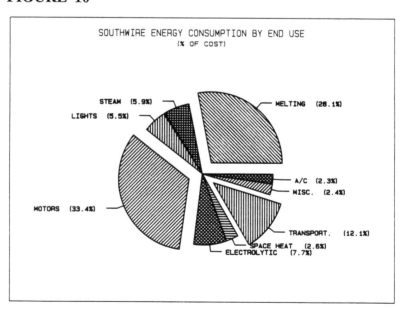

APPENDIX B 217

FIGURE 11

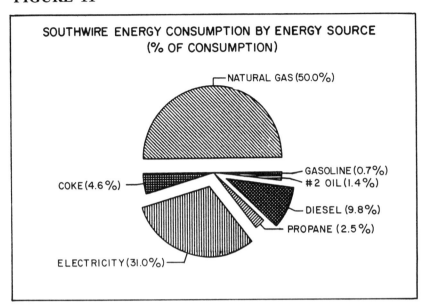

FIGURE 12

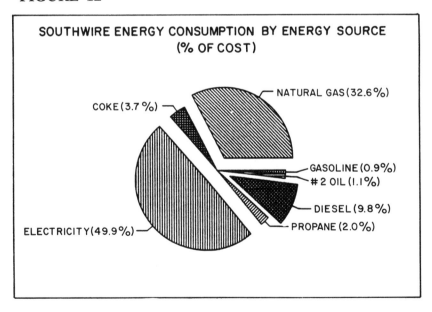

there will be between lower energy consumption and lower energy bills.

Vulnerability To Price Increases - The area of most vulnerability to price increase is in electrical usage, because those uses cannot be readily switched to other sources of energy. As Figure 12 shows, electricity accounts for about 50% of energy cost. Thus, any increase in electricity cost has a large impact on Southwire's cost of energy.

On the positive side, electricity price increases can be mitigated somewhat by controlling demand and negotiating rates with the utility companies.

With respect to nearly all other fuel usage, we do have the capability to use alternate fuels and thus lessen the impact of a price increase for a given fuel. Nearly all equipment using natural gas, including furnaces and boilers, can be quickly switched to #2 oil or propane, and in fact this is done several times a year when natural gas is curtailed. Given the ease of switching fuels for most of the equipment, the major concern becomes one of maintaining adequate #2 oil and propane supplies, and this is being done.

We recently completed a test program using compressed natural gas (CNG) to power forklifts. This program demonstrated that, should propane prices increase significantly in the future, we could convert the forklifts to run on natural gas.

Negotiating Strength With Utilities - During the last two years, we have become more active in negotiating rates with power and gas companies. The actions taken included: intervening in gas and electricity rate cases; developing contacts in the gas and electric companies to whom we can go with our questions and concerns; directly negotiating rates with utilities and proposing new rates; actively discussing cogeneration projects with the electric utilities; purchasing gas directly from producers; and purchasing gas at prices competitive with #5 fuel oil.

In general the utility companies have become much more willing to negotiate with Southwire in the last few years. This trend will continue as the utilities face more competition from other suppliers and industry.

Demand Control - Since almost one-half of Southwire's electricity bill is due to demand charges, demand control can significantly reduce the bill. The energy management system is a large step in that direction, as are the peaking engines.

Problems still remain in the area of demand control, however. Coordination with the plant operating personnel can be improved, so that when we receive notice to cut demand in the summer, we will get the full cooperation of all the plants. This must be improved to obtain the maximum possible demand savings.

Another aspect of demand control which needs improvement is the identification of the proper machines to turn off so as to minimize the impact on plant operations. Again, this will require the input and support of personnel in the various plants.

Demand control equipment is in place at all manufacturing plants. We work constantly with our suppliers to develop rates that allow savings from demand control.

Relations With Other Departments - As mentioned earlier, the most effective way to reduce energy consumption can be to assign responsibility for it to plant personnel. This also involves obtaining the cooperation of all personnel with respect to energy programs. This is an especially important area at Southwire because the Energy Management department is on the corporate and not plant level. This situation can lead to an "us vs. them" mentality, seriously hampering the effectiveness of any energy programs. It is crucial to avoid giving plant personnel the idea that Corporate Energy Management is trying to force projects on them.

Another area of potential conflict is in capital budgeting. Understandably, those in operations are primarily concerned with production, and often see expenditures on energy projects as taking money from badly-needed production equipment. However, from the company's standpoint a dollar of energy saved is just as valuable as a dollar saved in another area, or an additional dollar of revenue. Therefore, energy projects sould be given a fair chance to compete with other projects for capital funds. In fact, energy projects can be more attractive than production-related projects, because of their more certain returns. With regard to capital budgeting, then, we also need to get corporate and production people "on our side" and avoid an adversarial relationship.

We have been making progress on better relationships with groups within Southwire. We must work closely with plant managers to have energy projects and programs included in their capital and operating budgets.

Another means of obtaining plant support is to conduct special energy-related programs. In the past, this has been done with Energy Awareness Month, and various task forces, such as the Motor Task Force. These programs have been helpful in increasing energy awareness, but should be intensified in the future.

Implementation Of New Technology - New technology can be very important in reducing the energy component of manufacturing cost, because in many cases it consumes considerably less energy than the technology it replaces. One of the key functions of the Corporate Energy Management department is to evaluate the energy saving capability of new technology, the applicability of the technology to Southwire, and the economics of the technology. Applicability is especially important because certain technology may perform well at other plants but not at Southwire due to differences in plant operations. With new technology, once it is established that it is applicable to our operation, the bottom line is payback or return on investment. That is, will the energy savings pay back the capital cost in a reasonable period of time?

There is great potential for new energy technology at Southwire in areas such as steam production and distribution, compressed air, melting, smelting, incineration, and alternate fuels.

As Figures 9 and 10 show, melting, motors, transportation, steam, and electrolytic refining account for 89% of energy consumption and 87% of energy cost. In these areas new technology is especially important because it can provide "quantum" reductions in energy usage rather than small changes.

New technologies we will explore in the years ahead include:
- Oxygen trim controls on boilers and aluminum furnaces
- Ceramic recuperators
- Oxygen fuel burners
- Oxygen and nitrogen production
- Wood/coal power plant
- Combined cycle generation
- Gas turbine cogeneration with steam recovery
- PVC incineration
- Industrial heat pumps
- Vapor recompression
- Baseload engines with heat recovery
- Two stage absorption air conditioning
- Substitution for ammonia in copper refining
- More automated burner management
- Co-fired boiler burners
- Upgraded compressed air systems

APPENDIX C

PORTABLE ENERGY AUDIT INSTRUMENTS

SLING PSYCHROMETER

Measures: Wet and dry bulb temperatures to obtain relative humidity.

Uses: ·Check operation of heating/air conditioning systems - compare inside to outside relative humidity.

·Check humidity conditioning in warehouses, etc.

Range: 300 to 1100F

Approx. Cost: $45

LIGHT METER

Measures: Levels of Illumination measured in Foot Candles.

Uses: Comparison of actual lighting levels to standards, design levels or actual requirements.

Range: 0-200 foot candles, color and cosine corrected.

Approx. Cost: $45 - $400

MULTIMETER

Measures: Volts, amps, resistance, conductance and surface temperatures.

Uses: ·Motor load survey (Amp reading vs. nameplate amps)

·Voltage/voltage fluctuations.

Approx. Cost: $270

DIAL THERMOMETERS

Measures: Exhaust temperatures

Uses: Stack gas temperatures on boilers, steam generators, ovens, etc.

Range: 2000-10000 F

Approx. Cost: $30

SURFACE TEMPERATURE THERMOMETERS

Measures: Actual/maximum/minimum surface temperatures

Uses: Surface temperatures of pipes, ovens, furnace walls, boiler surfaces, motors, air ducts, freezer walls, etc.

Range: 500 -10000 F

DIGITAL THERMOMETER WITH PROBES

Measures: Temperatures - surfaces, liquid, air, exhaust, etc.

Uses: Surface temperature probe

·Steam traps

·Steam pipes/pipe insulation

·Product lines

·Tank surface

·Motor operating temperatures

·Motor bearings

·Circuit breakers, electrical parts
General Temperature probe
·Liquids
·Air (heated, cooled)
·Exhaust

Range: 00 - 9990 F
Approx. Cost: $350

AIR VELOCITY METER

Measures: Air velocity (fpm) and temperature of air flow
Uses: ·Flow measurements, i.e., ventilation fans
·Velocity profiles
·Monitoring outdoor air movements
·Balancing/trouble shooting HVAC systems
·Engine cooling systems

Range: 0 - 6000 fpm, 00 -2000 F
Approx. Cost: $700

ULTRASONIC LEAK DETECTOR

Measures: Air borne internally generated ultrasounds
Uses: ·Leak detection - pressure, vacuum, steam, compressed air, fuel, hydraulic, etc.
·Steam trap malfunctions
·Valve leakage/line blockage

·Motor bearing checks

·Compressor heads

·Missing teeth on gears

Approx. Cost: $800

O2 and CO2 FYRITE STACK GAS ANALYZER

Measures: O_2 and CO_2 content in flue gas

Uses: Determines combustion efficiency of boilers, furnaces, etc. (Must measure stack temperature at same time)

Approx. Cost: $200

HAND-HELD NON-CONTACT INFRARED THERMOMETER

Measures: Surface temperatures

Uses: Measure surface temperatures of:

·Walls

·Pipes

·Steam traps

·Boilers

·Furnaces

·Moving objects

·Castings

Approx. Cost: $1800

APPENDIX D
A SAMPLE AUDIT
ENERGY AUDIT RECOMMENDATIONS

Priority "A" - immediate action
Priority "B" - 1982 implementation (in 1985 budget)
Priority "C" - long range

ITEM	AREA	UNIT/SYSTEM	RECOMMENDATIONS	ESTI. INST. COST	NET ESTI. SAVINGS	PAYBACK (MOS.)	PRIORITY
1.	Utilities	General Instrumentation Calibration	Repair/calibrate all existing instrumentation on chillers, boilers and air compressors for accurate/credible monitoring. Field zero and span once/6-12 months according to a new priority list based on critical needs.	$12,000(E) (12 man wks plus mat'ls)	$12,000 (0.2% of $6MM purchased energy bill)	12	A
2.	Utilities	Energy Efficient Instrumentation and Monitoring Study and Hardware Upgrade	Commit to a study of on-line process monitoring and control requirements. Retrofit for all major utility systems, including boilers, air compressors, cooling towers and refrigeration machines. Add additional monitoring instrumentation with microprocessor base to allow complete unit energy balances and performance checks. Provide on-line unit efficiency values (KW/Ton, BTU fuel/lb. steam exported, KWH/CFM, % vented, etc.)	$40,000(E) Purchased Study 240,000(C) microprocessor based instrumentation (100 primary sensors at $2400 each)	$120,000 (2.0% of $6MM purchased energy bill)	28	B
3.	Utilities	Chilled Water Distribution System	Repair wet insulation on large (20") chilled water main header located in vertical section at back of 415 corridor.	$3,000(E) (20 man hours plus materials)	$1,000 (Reduction in heat gain to chilled water of 0.01°F)	36	A

227

228 MANAGING ENERGY RESOURCES IN TIMES OF DYNAMIC CHANGE

ITEM	AREA	UNIT/SYSTEM	RECOMMENDATIONS	ESTI. INST. COST	NET ESTI. SAVINGS	PAYBACK (MOS.)	PRIORITY
4.	Utilities	Chilled Water Distribution System	Investigate economics and pursue, if justified, construction of underground chilled water storage to be used for demand control, emergency back up, and to delay purchase of new chiller capacity.	$300,000(C) (2 hours of 3% chilled water storage at 20,000 GPM, 40'x40'x200' covered pit with pumps.	$100,000 ($36,000 per year in reduced demand and $64,000 per year lost mfg. margin due to chilled water interruption or peak capacity problems.	36	C
5.	Utilities	Air Compressors	Investigate and implement control system modifications with vendor to match system output with demand and thereby minimize venting. Provide sequential modulating control on all compressors.	$60,000(C) (2 systems, $40,000 for 6 low pressure units, $20,000 for 4-hour press units)	$80,000 Reduce venting from 15% to 5%.	16	B
6.	Utilities	Air Compressor Filters	Change air inlet filters on most compressors. Install magnahelic DP gauge and establish optimum filter change frequency for minimum energy consumption.	$5,000(C) (10 compressors $500 each)	$4,000 (Reduce compressed air power cost by 0.5%)	15	A
7.	Utilities	Plant Air and Steam Flow Measurement and Accountability	Install compressed air and steam flow integrators for plant areas. Establish consumption goals for various production levels and promote conservation as an important and necessary part of the job.	$36,000(C) (18 meter stations at 2000 each)	$16,000 (Reduced consumption by 2.0%)	27	B
8.	Utilities	Compressed Air Distribution System	Install air condensate traps with sight glasses on aftercoolers. Do not bypass traps as a routine procedure.	$6,000(C) (12 locations at $500 each)	$4,000/yr (Reduce consumption by 0.5%)	18	A

APPENDIX D 229

ITEM	AREA	UNIT/SYSTEM	RECOMMENDATIONS	ESTI. INST. COST	NET ESTI. SAVINGS	PAYBACK (MOS.)	PRIORITY
9.	Utilities	Boilers Blowdown	Recalibrate, tune level controller on continuous blowdown tank to use proportional band and eliminate reset	$500 (E)	$500 (1.0% reduction in blowdown chemicals, etc.)	12	A
10.	Utilities	Boiler Fuel Oil Supply	Lower recirculation temperature of fuel oil from existing 200°F and consider shutting down booster heater pending evaluation of heat up response time	$100 (E)	$1,000 (700 pph flow rate with 30°F DT loss)	1	A
11.	Utilities	Boiler Deaerator System	Shut off steam tracing on deaerator piping during summer months	$100 (E)	$2,000 (50 pph reduction for 9 mos.)	1	A
12.	Utilities	Boiler Combustion	Initiate immediate manual monitoring of excess oxygen on daily or shift basis with portable analyzer until continuous on-line oxygen analysis and automatic trim is available.	$3,000(E) $3,000(C)	$4,000 (1% reduction in $1.6MM fuel bill, 3 mo.)	18	A
13.	Plant-wide	Steam Traps	Revitalize steam trap program initiated in 1980. Systematically inspect and repair 1200 active traps with full time attention on a specified schedule	$40,000(E)	$80,000	6	A
14.	Plant-wide	Steam Leaks	Continue with outside leak repair service for routine maintenance to insure tight steam distribution system	$12,000	$48,000	3	A
15.	Utilities	Boilers	Install two new packaged coal fired boilers (20,000 pph each) using existing fuel oil-natural gas boilers as emergency standby	$3.0MM(C)	$1.047MM 1983 net cost reduction	34	C
16.	Plant-wide	Solar Evaporative Roof Cooling	Install solar evaporative roof cooling to reduce direct solar roof load above north and south drawtwist areas (411, 412).	$20,000(C) (20,000 ft² at 35/ft² plus water supply, etc.)	$11,500 (100 ton reduction in load, 3 mo/yr, 12 hr/day $64/MTH plus demand	21	B

ITEM	AREA	UNIT/SYSTEM	RECOMMENDATIONS	ESTI. INST. COST	NET ESTI. SAVINGS	PAYBACK (MOS.)	PRIORITY
17.	Staple Area	Air Balance in Spinning	Reinstall airlock doors and take-up air back to spin floor and quench air units using super-sucker with (1) new pressure control or (2) directly into existing returns	$20,000(E)	$6,000 (20,000 CFM with 6°FT)	40	B
18.	Spin Area	Lighting	Add switches to existing lighting behind spin machines on take-up floor	$1,000(C)	$500 (10 HO fixtures $50/yr. ea.)	24	B
19.	Spinning	Lighting in Blast Chamber	Install and use light switches on 16 fixtures (455 watt each).	$500(C)	$3,200 (8 VHO fixture, $100/yr. elec, $100/yr. AC)	2	A
20.	Texturizer Area	Lighting	Install one light fixtre per texturing machine, positioned as task lighting at lower level versus existing general area lighting. Full implementation would eliminate 80 of existing 200 fixtures.	$40,000(E) (Relocate, etc. 200 fixtures, $200 each)	$16,000 (Elim. 80 Fixtures at $200/yr. ea.)	30	B
21.	Waste Treat-	Spin Finish Evaporators (Active Project)	Determine most economical means to concentrate waste spin finish	$200,000(C)	$158,000 (80% reduction on 5000 pph 75% on Time)	15	B
22.	Storage Area	Lighting	Add and use light switch for outside lighting (lights on during days)	$500 (E)	$220 (save 2 KW 50% Time)	27	A
23.	Condensation 421, 422 Areas	Exhaust Fan	Install thermostat switch on exhaust fans on roof.	$4,000(C) (10 fans, 10 HP each)	$8,000 (40% shut off)	6	A
24.	Maint, Engr, Bldg, 403	HVAC	Install 365 day programmable time clock to control AC loads, lights, etc. Control AC on 85°F summer reset and 55°F winter set back. Currently conditioning building 24 hrs/day, d 7 days/week.	$10,000(C)	$18,000 (save 50% of 3000/mo., purchase, e.g. bill)	7	A

APPENDIX D 231

ITEM	AREA	UNIT/SYSTEM	RECOMMENDATIONS	ESTI. INST. COST	NET ESTI. SAVINGS	PAYBACK (MOS.)	PRIORITY
25.	Warehouse Area	Bump Doors	Close bump doors between bale storage and packout areas to reduce air exfiltration.	$100(E)	$330 exfiltration with 15° T half time	3.6 1000 CFM	A
26.	A Bldg.	Programmable Time Clock Controller	Install a 365 day programmable time clock controller on HVAC and lighting. Discontinue present practice of cooling and heating entire building on 2nd, 3rd, shifts and weekends.	$10,000(C)	$9,000 (25% savings in $3,000/mo. energy bill)	13	B
27.	Plant-wide	Energy Management Program	The overall Plant Energy Management Program should include the following elements:				
A.		Top Management	Formally notify all employees that the energy management program has the complete support of top management. This might be best accomplished with a letter from the General Manager outlining the importance of energy conservation to our jobs and our future.	$1,000(E)			
B.		Plant Energy Committee	Reactivate an energy committee with emphasis on area responsibility for monitoring consumption/performance, promoting conservation, initiating area energy-related improvements and implementing operational changes resulting in energy savings. Monthly meetings should be brief and according to a published agenda. Committee activity must have the full support of top management.	$10,000(E)			
C.		Plant Energy Planning	Issue and update annually a rolling five (5) energy plan. The plan should include a major capital expense schedule by individual projects and a contingency plan to be used in the event of an extended fuel or electrical power outage.	$4,000(E)			
D.		Energy Reporting	Issue monthly reports covering (1) current monthly energy accomplishments and immediate plans, (2) monthly energy consumption and performance, and (3) annual report highlighting accomplishments for calendar year.	$6,000(E)			

ITEM	AREA	UNIT/SYSTEM	RECOMMENDATIONS	ESTI. INST. COST	NET ESTI. SAVINGS	PAYBACK (MOS.)	PRIORITY
	E.	Employee Conservation Involvement	Continue promotion and education with employees on the value of conservation and individual responsibility (energy computer, etc.). Promote energy cost conscientiousness with energy unit cost data and with energy operating cost information actually posted on equipment.	$4,000 (E)			
	F.	Area or Machine Energy Balance Analysis	Perform approximate energy balances on all major energy consuming areas or units. Compare actual energy efficiencies with maximum practical efficiencies to determine avoidable waste.	$20,000(E)			
	G.	Survey Audits	1. Perform "twenty minute" walk-through audits in each plant area by a three-man team once every two months to assess energy conservation performance.	$2,000(E) 3 man hrs/audit 6 plant areas 6 times/year $20/man hour			
			2. Perform a 3-5 day short audit once per year by an internal audit team.	$6,000(E)			
			3. Perform an intensive 10-15 day audit once every 3 years with the assistance of an outside consultant.	$6,000/yr(E)			
		TOTAL	ENERGY PROGRAM ADMINISTRATION	$59,000			

Appendix E

SOUTHWIRE'S STRATEGIC ENERGY PLAN[1]

John T. Kopfle, P.E., C.E.M.
Senior Energy Engineer
Corporate Energy Management Department
Southwire Company
P.O. Box 1000
Carrollton, Georgia 30119-0001

BACKGROUND

Southwire Company, Carrollton, Georgia, manufactures copper and aluminum wire and cable for commercial and residential buildings, utilities, electronic manufacturers, and others. The company operates 10 plants in 5 states and employs over 4,000 people. In addition to its wire and cable plants, Southwire owns an electrolytic copper refinery, a machinery division, an automated sawmill and a hydroelectric plant.

Southwire's Corporate Energy Management (CEM) Department has the task of reducing the energy component of manufacturing costs and assuring adequate supplies of energy for all the facilities mentioned above. The CEM Department's activities include:
1. Compiling energy consumption data for all plants and billing them monthly.
2. Operating energy properties such as a hydroelectric plant and peaking engine generators.
3. Operating Energy Management Systems (EMS) to reduce electrical demand charges.

---

[1] This is an updated version of an article published in Strategic Planning and Energy Management, Vol. 5, No. 1, 1985, p.5

4. Identifying and developing major capital projects which will yield a favorable return while reducing energy cost and/or improving reliability.

One indication of the success of Southwire's Energy program is that from 1981-86 energy used per pound of product decreased by almost 40%. This has resulted in a savings to the company of about $37 million over that period. In addition, cost avoidance savings over this period have been substantial.

With the growth and success of Southwire's energy management program, the department has recognized the need for more formal planning. To meet that need, CEM has developed a strategic energy plan.

THE STRATEGIC PLANNING PROCESS

To formulate a strategic energy plan at southwire we first compiled data on energy usage by both end use and energy source. By doing so, we had a better idea of the areas which were using the most energy and which should get the most attention in the strategic plan.

For several years, we have been compiling data on energy consumption and storing it in the company's mainframe computer. This greatly facilitated gathering the data needed for the strategic plan. It was necessary to break out consumption by end use (for example motors, lighting, air conditioning, etc.), and this was accomplished by gathering energy audit data and making inquiries of plant operating personnel.

The results of the study are not meant to be exact, but do give a good breakdown of energy use. Four separate charts were developed, which are shown in Figures 1-4. Pie charts are especially useful for this type of information.

Energy cost, as well as BTU's consumed, should be considered when developing these types of charts. After all, the heart of the mission of a CEM department is saving money, not necessarily energy. This is a point that can be overlooked in the zeal to conserve energy.

A wealth of information can be gleaned from these figures. Consider, for example, lighting, space heating, and air conditioning. For Southwire these costs constitute a relatively small percentage of the total energy consumed. This is primarily because Southwire is an industrial facility, with large process energy requirements. An office complex, hospital, or college, however, would likely have most of its energy consumption going toward lighting and space conditioning. By developing breakdowns of energy usage such as these, one can begin to see where the CEM department's effort should be placed.

After compiling this data, we began the actual planning process. The basic idea of strategic energy planning is to look at the energy situation facing the company, discover any potential energy-related problems which may occur, and determine how the company will be able to cope with them. Then, plans are formulated to deal with the vulnerable areas. Obviously, this process must be tailored to each company's situation, since no two companies are alike. For this reason, there is a danger in relying too heavily on a consulting firm to formulate a company's strategic energy plan. Due to the scope and importance of the effort, an intimate knowledge of the company's unique energy situation is essential.

The format we used for the strategic energy plan is based on one described in an issue of Strategic Planning and Energy Management[2].

---

[2] Cooper, R. H.: "Energy Strategic Planning" Strategic Planning & Energy Management Vol. 4 No. 2, 1984, p.20

The format is as follows:

I. EXECUTIVE SUMMARY
II. SITUATION ANALYSIS
    A. External Factors
    B. Internal Strengths and Weaknesses
III. ENERGY REQUIREMENTS AND COSTS
IV. KEY ISSUES, OBJECTIVES, AND STRATEGIES
V. ACTION PLANS
VI. PLAN RESULTS

Perhaps the most important activity in strategic planning is developing key issues, those matters which would seriously affect the company's operations if not resolved. For that reason, and for the sake of brevity, I will focus on the key issues which we have developed and the objectives, strategies and actio plans which we will use to deal with them.

KEY ISSUE #1 - ENERGY AS A PERCENT OF CONVERSION COST

Although it constitutes a small percentage of cost of sales, energy is a significant component of Southwire's conversion cost. Conversion cost is more significant in this case than cost of sales because conversion cost includes those costs (energy, labor, maintenance) which are somewhat controllable, whereas a large part of cost of sales is material cost, which is largely uncontrollable. Thus, control of energy cost can contribute significantly to profitability.

Objectives

For this key issue, we have established two objectives, one related to energy conservation and one related to cost avoidance:

1. To reduce energy consumption per pound of product at Southwire manufacturing facilities by 5% per year and reduce overhead energy consumption by 2% per year, over the next three years.

2. To implement applicable cost avoidance measures to reduce Southwire's energy cost.

Strategies

Conservation — The two primary means of reducing the amount of energy used are revising operating practices and implementing new technology.

Cost Avoidance — This involves minimizing the cost per unit of energy purchased, rather than the amount of energy used. Some of the means are: using energy management systems to control electrical demand charges; agressively negotiating the lowest possible energy rates; purchasing natural gas directly; and retaining flexibility in fuel use.

KEY ISSUES #2 - HEAVY RELIANCE ON SINGLE FUEL SOURCES

When the oil crises of 1973-74 and 1979 occurred and companies began to take energy seriously, the major consideration was reliability of supply. This situation has changed in recent years, and the primary concern is now price. Nevertheless, heavy reliance on one fuel source for a particular end use can cause serious problems if that fuel source is interrupted. Even short-lived interruptions can be costly if a critical process is involved.

Objectives

For this key issue, we have developed two objectives:
1. Insofar as possible, ensure reliability of supply for all energy sources.
2. Where necessary, develop flexibility in energy use.

Strategies

Reliability — To assure reliability, these steps should be taken: for those fuels which require storage, maintain adequate onsite storage;

	maintain plant energy delivery systems in good condition and make sure their capacity is sufficient; and maintain energy using equipment in good condition.
Flexibility	- Since perfect reliability cannot be assured, flexibility in energy use is desirable, so that if one energy source is unavailable, another can be substituted. The primary strategy for maintaining flexibility is to ensure that all equipment, insofar as possible, has the ability to operate on alternate fuels.

KEY ISSUE #3 - ELECTRICITY RATE SHOCK

A third key issue for Southwire is the large electricity rate increases caused by the power plants coming on line in Georgia, Kentucky, Arkansas, and Connecticut. These types of rate increases are known as "rate shock". There are two reasons for the importance of this issue: first, as Figures 3 and 4 show, electricity accounts for about one third of Southwire's energy consumption and one half of energy cost; secondly, most uses of electricity cannot switch readily to other energy sources. The projected rate increases also will have a significant impact on Southwire's cost of sales.

Objective

For the key issue of electricity rate increases, our objective is to minimize the impact of these increases on Southwire's electricity costs over the next five years.

Strategies

To minimize increases in Southwire's electricity costs, we will pursue four general strategies. Some of these strategies are similar to the strategies developed for the first two key issues. However, due to the importance of this key issue, we felt it was necessary to reemphasize them.

1. Conservation - reducing the amount of electricity used is the most obvious strategy. Most of the

conservation savings we have realized to date have come from energy sources other than electricity. In particular, natural gas usage in melting applications has been decreased significantly. Reductions in electricity use should receive more emphasis from now on, however, because of our vulnerability to electricity price increases, as discussed above.

There are two basic ways to reduce electrical consumption: revise operating practices and implement new technology. Some specific examples are given in the Action Plans section.

2. Load Management - this involves managing electrical loads so as to minimize demand charges. Thus, although no energy is saved, costs are avoided. This is carried out by the use of our energy management system and peaking engine generators.

3. Favorable Utility Policies - utilities have become much more willing to negotiate in recent years, and Southwire plans to take full advantage of this situation. In areas such as sales contracts, backup rates, and wheeling, the company will make sure its position is heard.

4. Self Generation - in cases where the company is unable to negotiate favorable rates, generating our own power becomes a viable option. Southwire is already doing this with a hydroelectric plant, but there may be additional such opportunities in the future.

ACTION PLANS

To keep this paper to a reasonable size, it is not possible to lay out action plans in great detail. However, action plans can be listed in general terms to provide a basis for the CEM department's activities for the next five years.

STRATEGY		ACTION PLANS
Conservation	Electricity	-continue "house cleaning" efforts
	Natural Gas	-maintain and operate burners properly
	Fuel Oil	-maintain and operate burners properly
	Propane	-maintain and operate burners properly
		-operate LP/air plants at proper mix
		-use up-to-date burner controls

STRATEGY		ACTION PLANS
Conservation	All Energy Sources	-optimize end use conservation -improve metering and billing system to place responsibility for conservation on end users
	All Energy Sources	-increase plant energy programs such as BTU boards and task forces to improve relations with plant operations personnel
	Lights	-use high efficiency lights -use skylights and photo cells where feasible -develop improved layouts and standards
	Motors	-replace defective motors with energy efficiency types, rather than rewinding -use AC inverters for speed control where feasible -use motors with proper size and speed for given application
	Compressed Air	-perform periodic leak surveys -substitute electronic for air control -use high efficiency compressors -reduce unnecessary use
	Steam	-perform periodic leak surveys -install better insulation -use proper fittings and valves -install drip traps where needed -manage use to end demand spikes
	Metal Melting	-investigate use of oxygen burners -use high velocity burners -use heat recovery where feasible -optimize furnace controls -change process cycle to reduce energy requirements
Cost Avoidance	Electricity	-use energy management system (EMS) to control demand -work with plant personnel to obtain better cooperation during demand control periods -supplement EMS with operation of peaking engine generators -obtain load management power rates -support competitive cost rate making
	Natural Gas	-use interruptible service where feasible -stay abreast of natural gas prices and selling networks and arrange for "self help" gas where possible -support competition in the gas industry
	Fuel Oil	-have alternate suppliers available -store only for seasonal needs
	Propane	-purchase in bulk
Reliability	Electricity	-maintain adequate plant electrical systems -urge adequate supplier facilities

STRATEGY		ACTION PLANS
	Natural Gas	-support deregulation of gas industry to ensure continuous supply
	Fuel Oil	-maintain adequate onsite storage -ensure adequate supplier capability
	Propane	-use heavy duty vaporizing and mixing equipment for LP/air plants
Flexibility	Electricity	-use self generation equipment where feasible, using parallel generation with black start ability -maintain load shifting ability (time of day) -urge utility competition and wheeling -obtain cooperation of plant personnel to identify new areas where demand control can be used
	Natural Gas	-maintain standby fuel storage and alternate fuel burning equipment -investigate means to give CDS and CRD shaft furnaces ability to burn alternate fuels easily
	Coke	-consider use of oxygen burners as alternative -develop ability to use lower grade coal -maintain supplier competition
	Fuel Oil	-maintain ability to use #6 as well as #2 oil
	Propane	-maintain oil firing ability where possible
Favorable Utility Policies	Electricity	-negotiate with power companies to obtain lowest possible rates
Self Generation		-continue assessments of feasibility of cogeneration and small power projects -investigate financing options

IMPLEMENTATION

The most difficult task in strategic energy planning is implementing the plan so that it affects the day-to-day operation of the company. This is accomplished in part by using the plan as an input for other planning activities at Southwire. Both the overall company business plan and individual plant planning efforts rely on the Strategic Energy Plan to determine the way in which energy issues will affect the company's operations. Further, this ensures energy consciousness throughout Southwire.

We also are using the plan to provide the basis for assignments in the CEM Department. The action plans from the Strategic Energy Plan become objectives for individuals in the department. The individuals then develop strategies and action plans to accomplish these objectives. In this way, we are certain that each part of the plan becomes the responsibility of a certain individual.

CONCLUSION

The use of Strategic Energy Planning enables Southwire to cope with energy-related problems which have a high probability of occurring in the future. The first step was to assess the current situation, both the external energy environment and outhwire's strengths and weaknesses with respect to energy. From these assessments we identified key issues and potential problem situations which could have a detrimental effect on Southwire's long-term profitability. The three issues identified were:

1. Energy as a percent of conversion cost;
2. Heavy reliance on single fuel sources;
3. Electricity rate shock.

We have developed objectives, strategies, and action plans to deal with each of the key issues.

Southwire's strong commitment to energy management has resulted in a successful program thus far. Nevertheless, there is still much potential for further savings. The Corporate Energy Management Department strongly believes that Strategic Energy Planning is providing for more effective planning and management of the energy program. It is also preparing the company to deal more effectively with serious energy related problems which are likely to occur. By identifying these potential problems early enough, CEM can develop plans and gather resources to cope with them, and thereby minimize the deleterious effect on Southwire's financial performance.

Figure 1

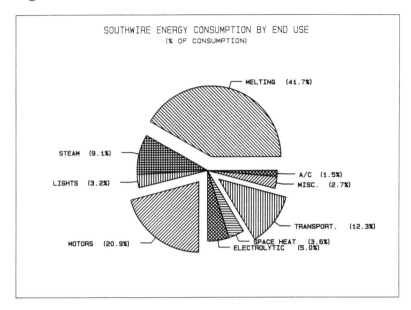

Figure 2

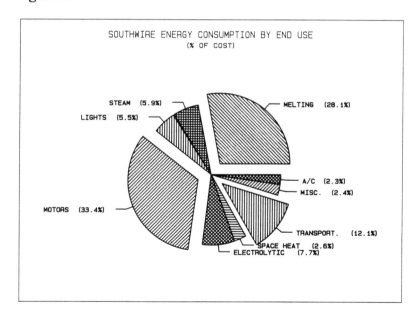

Figure 3

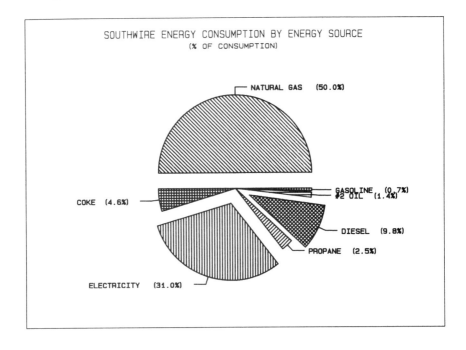

Figure 4

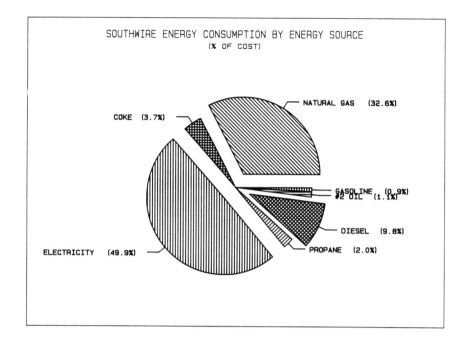

Appendix F

2% COMPOUND INTEREST FACTORS

	Single Payment		Uniform Series			
	Compound Amount Factor	Present Worth Factor	Capital Recovery Factor	Present Worth Factor	Sinking Fund Factor	Compound Amount Factor
Periods n	F/P	P/F	A/P	P/A	A/F	F/A
1	1.02000	.980392	1.02000	.980392	1.00000	1.00000
2	1.04040	.961169	.515050	1.94156	.495050	2.02000
3	1.06121	.942322	.346755	2.88388	.326755	3.06040
4	1.08243	.923845	.262624	3.80773	.242624	4.12161
5	1.10408	.905731	.212158	4.71346	.192158	5.20404
6	1.12616	.887971	.178526	5.60143	.158526	6.30812
7	1.14869	.870560	.154512	6.47199	.134512	7.43428
8	1.17166	.853490	.136510	7.32548	.116510	8.58297
9	1.19509	.836755	.122515	8.16224	.102515	9.75463
10	1.21899	.820348	.111327	8.98259	.091327	10.9497
11	1.24337	.804263	.102178	9.78685	.082178	12.1687
12	1.26824	.788493	.094560	10.5753	.074560	13.4121
13	1.29361	.773033	.088118	11.3484	.068118	14.6803
14	1.31948	.757875	.082602	12.1062	.062602	15.9739
15	1.34587	.743015	.077825	12.8493	.057825	17.2934
16	1.37279	.728446	.073650	13.5777	.053650	18.6393
17	1.40024	.714163	.069970	14.2919	.049970	20.0121
18	1.42825	.700159	.066702	14.9920	.046702	21.4123
19	1.45681	.686431	.063782	15.6785	.043782	22.8406
20	1.48595	.672971	.061157	16.3514	.041157	24.2974
21	1.51567	.659776	.058785	17.0112	.038785	25.7833
22	1.54598	.646839	.056631	17.6580	.036631	27.2990
23	1.57690	.634156	.054668	18.2922	.034668	28.8450
24	1.60844	.621721	.052871	18.9139	.032871	30.4219
25	1.64061	.609531	.051220	19.5235	.031220	32.0303
26	1.67342	.597579	.049699	20.1210	.029699	33.6709
27	1.70689	.585862	.048293	20.7069	.028293	35.3443
28	1.74102	.574375	.046990	21.2813	.026990	37.0512
29	1.77584	.563112	.045778	21.8444	.025778	38.7922
30	1.81136	.552071	.044650	22.3965	.024650	40.5681
31	1.84759	.541246	.043596	22.9377	.023596	42.3794
32	1.88454	.530633	.042611	23.4683	.022611	44.2270
33	1.92223	.520229	.041687	23.9886	.021687	46.1116
34	1.96068	.510028	.040819	24.4986	.020819	48.0338
35	1.99989	.500028	.040002	24.9986	.020002	49.9945

3% COMPOUND INTEREST FACTORS

	Single Payment		Uniform Series			
	Compound Amount Factor	Present Worth Factor	Capital Recovery Factor	Present Worth Factor	Sinking Fund Factor	Compound Amount Factor
Periods n	F/P	P/F	A/P	P/A	A/F	F/A
1	1.03000	.970874	1.03000	.970874	1.00000	1.00000
2	1.06090	.942596	.522611	1.91347	.492611	2.03000
3	1.09273	.915142	.353530	2.82861	.323530	3.09090
4	1.12551	.888487	.269027	3.71710	.239027	4.18363
5	1.15927	.862609	.218355	4.57971	.188355	5.30914
6	1.19405	.837484	.184598	5.41719	.154598	6.46841
7	1.22987	.813092	.160506	6.23028	.130506	7.66246
8	1.26677	.789409	.142456	7.01969	.112456	8.89234
9	1.30477	.766417	.128434	7.78611	.098434	10.1591
10	1.34392	.744094	.117231	8.53020	.087231	11.4639
11	1.38423	.722421	.108077	9.25262	.078077	12.8078
12	1.42576	.701380	.100462	9.95400	.070462	14.1920
13	1.46853	.680951	.094030	10.6350	.064030	15.6178
14	1.51259	.661118	.088526	11.2961	.058526	17.0863
15	1.55797	.641862	.083767	11.9379	.053767	18.5989
16	1.60471	.623167	.079611	12.5611	.049611	20.1569
17	1.65285	.605016	.075953	13.1661	.045953	21.7616
18	1.70243	.587395	.072709	13.7535	.042709	23.4144
19	1.75351	.570286	.069814	14.3238	.039814	25.1169
20	1.80611	.553676	.067216	14.8775	.037216	26.8704
21	1.86029	.537549	.064872	15.4150	.034872	28.6765
22	1.91610	.521893	.062747	15.9369	.032747	30.5368
23	1.97359	.506692	.060814	16.4436	.030814	32.4529
24	2.03279	.491934	.059047	16.9355	.029047	34.4265
25	2.09378	.477606	.057428	17.4131	.027428	36.4593
26	2.15659	.463695	.055938	17.8768	.025938	38.5530
27	2.22129	.450189	.054564	18.3270	.024564	40.7096
28	2.28793	.437077	.053293	18.7641	.023293	42.9309
29	2.35657	.424346	.052115	19.1885	.022115	45.2189
30	2.42726	.411987	.051019	19.6004	.021019	47.5754
31	2.50008	.399987	.049999	20.0004	.019999	50.0027
32	2.57508	.388337	.049047	20.3888	.019047	52.5028
33	2.65234	.377026	.048156	20.7658	.018156	55.0778
34	2.73191	.366045	.047322	21.1318	.017322	57.7302
35	2.81386	.355383	.046539	21.4872	.016539	60.4621

4% COMPOUND INTEREST FACTORS

	Single Payment		Uniform Series			
	Compound Amount Factor	Present Worth Factor	Capital Recovery Factor	Present Worth Factor	Sinking Fund Factor	Compound Amount Factor
Periods n	F/P	P/F	A/P	P/A	A/F	F/A
1	1.04000	.961538	1.04000	.961538	1.00000	1.00000
2	1.08160	.924556	.530196	1.88609	.490196	2.04000
3	1.12486	.888996	.360349	2.77509	.320349	3.12160
4	1.16986	.854804	.275490	3.62990	.235490	4.24646
5	1.21665	.821927	.224627	4.45182	.184627	5.41632
6	1.26532	.790315	.190762	5.24214	.150762	6.63298
7	1.31593	.759918	.166610	6.00205	.126610	7.89829
8	1.36857	.730690	.148528	6.73274	.108528	9.21423
9	1.42331	.702587	.134493	7.43533	.094493	10.5828
10	1.48024	.675564	.123291	8.11090	.083291	12.0061
11	1.53945	.649581	.114149	8.76048	.074149	13.4864
12	1.60103	.624597	.106552	9.38507	.066552	15.0258
13	1.66507	.600574	.100144	9.98565	.060144	16.6268
14	1.73168	.577475	.094669	10.5631	.054669	18.2919
15	1.80094	.555265	.089941	11.1184	.049941	20.0236
16	1.87298	.533908	.085820	11.6523	.045820	21.8245
17	1.94790	.513373	.082199	12.1657	.042199	23.6975
18	2.02582	.493628	.078993	12.6593	.038993	25.6454
19	2.10685	.474642	.076139	13.1339	.036139	27.6712
20	2.19112	.456387	.073582	13.5903	.033582	29.7781
21	2.27877	.438834	.071280	14.0292	.031280	31.9692
22	2.36992	.421955	.069199	14.4511	.029199	34.2480
23	2.46472	.405726	.067309	14.8568	.027309	36.6179
24	2.56330	.390121	.065587	15.2470	.025587	39.0826
25	2.66584	.375117	.064012	15.6221	.024012	41.6459
26	2.77247	.360689	.062567	15.9828	.022567	44.3117
27	2.88337	.346817	.061239	16.3296	.021239	47.0842
28	2.99870	.333477	.060013	16.6631	.020013	49.9676
29	3.11865	.320651	.058880	16.9837	.018880	52.9663
30	3.24340	.308319	.057830	17.2920	.017830	56.0849
31	3.37313	.296460	.056855	17.5885	.016855	59.3283
32	3.50806	.285058	.055949	17.8736	.015949	62.7015
33	3.64838	.274094	.055104	18.1476	.015104	66.2095
34	3.79432	.263552	.054315	18.4112	.014315	69.8579
35	3.94609	.253415	.053577	18.6646	.013577	73.6522

5% COMPOUND INTEREST FACTORS

	Single Payment		Uniform Series			
	Compound Amount Factor	Present Worth Factor	Capital Recovery Factor	Present Worth Factor	Sinking Fund Factor	Compound Amount Factor
Periods n	F/P	P/F	A/P	P/A	A/F	F/A
1	1.05000	.952381	1.05000	.952381	1.00000	1.00000
2	1.10250	.907029	.537805	1.85941	.487805	2.05000
3	1.15763	.863838	.367209	2.72325	.317209	3.15250
4	1.21551	.822702	.282012	3.54595	.232012	4.31012
5	1.27628	.783526	.230975	4.32948	.180975	5.52563
6	1.34010	.746215	.197017	5.07569	.147017	6.80191
7	1.40710	.710681	.172820	5.78637	.122820	8.14201
8	1.47746	.676839	.154722	6.46321	.104722	9.54911
9	1.55133	.644609	.140690	7.10782	.090690	11.0266
10	1.62889	.613913	.129505	7.72173	.079505	12.5779
11	1.71034	.584679	.120389	8.30641	.070389	14.2068
12	1.79586	.556837	.112825	8.86325	.062825	15.9171
13	1.88565	.530321	.106456	9.39357	.056456	17.7130
14	1.97993	.505068	.101024	9.89864	.051024	19.5986
15	2.07893	.481017	.096342	10.3797	.046342	21.5786
16	2.18287	.458112	.092270	10.8378	.042270	23.6575
17	2.29202	.436297	.088699	11.2741	.038699	25.8404
18	2.40662	.415521	.085546	11.6896	.035546	28.1324
19	2.52695	.395734	.082745	12.0853	.032745	30.5390
20	2.65330	.376889	.080243	12.4622	.030243	33.0660
21	2.78596	.358942	.077996	12.8212	.027996	35.7193
22	2.92526	.341850	.075971	13.1630	.025971	38.5052
23	3.07152	.325571	.074137	13.4886	.024137	41.4305
24	3.22510	.310068	.072471	13.7986	.022471	44.5020
25	3.38635	.295303	.070952	14.0939	.020952	47.7271
26	3.55567	.281241	.069564	14.3752	.019564	51.1135
27	3.73346	.267848	.068292	14.6430	.018292	54.6691
28	3.92013	.255094	.067123	14.8981	.017123	58.4026
29	4.11614	.242946	.066046	15.1411	.016046	62.3227
30	4.32194	.231377	.065051	15.3725	.015051	66.4388
31	4.53804	.220359	.064132	15.5928	.014132	70.7608
32	4.76494	.209866	.063280	15.8027	.013280	75.2988
33	5.00319	.199873	.062490	16.0025	.012490	80.0638
34	5.25335	.190355	.061755	16.1929	.011755	85.0670
35	5.51602	.181290	.061072	16.3742	.011072	90.3203

6% COMPOUND INTEREST FACTORS

	Single Payment		Uniform Series			
	Compound Amount Factor	Present Worth Factor	Capital Recovery Factor	Present Worth Factor	Sinking Fund Factor	Compound Amount Factor
Periods n	F/P	P/F	A/P	P/A	A/F	F/A
1	1.06000	.943396	1.06000	.943396	1.00000	1.00000
2	1.12360	.889996	.545437	1.83339	.485437	2.06000
3	1.19102	.839619	.374110	2.67301	.314110	3.18360
4	1.26248	.792094	.288591	3.46511	.228591	4.37462
5	1.33823	.747258	.237396	4.21236	.177396	5.63709
6	1.41852	.704961	.203363	4.91732	.143363	6.97532
7	1.50363	.665057	.179135	5.58238	.119135	8.39384
8	1.59385	.627412	.161036	6.20979	.101036	9.89747
9	1.68948	.591898	.147022	6.80169	.087022	11.4913
10	1.79085	.558395	.135868	7.36009	.075868	13.1808
11	1.89830	.526788	.126793	7.88687	.066793	14.9716
12	2.01220	.496969	.119277	8.38384	.059277	16.8699
13	2.13293	.468839	.112960	8.85268	.052960	18.8821
14	2.26090	.442301	.107585	9.29498	.047585	21.0151
15	2.39656	.417265	.102963	9.71225	.042963	23.2760
16	2.54035	.393646	.098952	10.1059	.038952	25.6725
17	2.69277	.371364	.095445	10.4773	.035445	28.2129
18	2.85434	.350344	.092357	10.8276	.032357	30.9057
19	3.02560	.330513	.089621	11.1581	.029621	33.7600
20	3.20714	.311805	.087185	11.4699	.027185	36.7856
21	3.39956	.294155	.085005	11.7641	.025005	39.9927
22	3.60354	.277505	.083046	12.0416	.023046	43.3923
23	3.81975	.261797	.081278	12.3034	.021278	46.9958
24	4.04893	.246979	.079679	12.5504	.019679	50.8156
25	4.29187	.232999	.078227	12.7834	.018227	54.8645
26	4.54938	.219810	.076904	13.0032	.016904	59.1564
27	4.82235	.207368	.075697	13.2105	.015697	63.7058
28	5.11169	.195630	.074593	13.4062	.014593	68.5281
29	5.41839	.184557	.073580	13.5907	.013580	73.6398
30	5.74349	.174110	.072649	13.7648	.012649	79.0582
31	6.08810	.164255	.071792	13.9291	.011792	84.8017
32	6.45339	.154957	.071002	14.0840	.011002	90.8898
33	6.84059	.146186	.070273	14.2302	.010273	97.3432
34	7.25103	.137912	.069598	14.3681	.009598	104.184
35	7.68609	.130105	.068974	14.4982	.008974	111.435

7% COMPOUND INTEREST FACTORS

	Single Payment		Uniform Series			
	Compound Amount Factor	Present Worth Factor	Capital Recovery Factor	Present Worth Factor	Sinking Fund Factor	Compound Amount Factor
Periods n	F/P	P/F	A/P	P/A	A/F	F/A
1	1.07000	.934579	1.07000	.934579	1.00000	1.00000
2	1.14490	.873439	.553092	1.80802	.483092	2.07000
3	1.22504	.816298	.381052	2.62432	.311052	3.21490
4	1.31080	.762895	.295228	3.38721	.225228	4.43994
5	1.40255	.712986	.243891	4.10020	.173891	5.75074
6	1.50073	.666342	.209796	4.76654	.139796	7.15329
7	1.60578	.622750	.185553	5.38929	.115553	8.65402
8	1.71819	.582009	.167468	5.97130	.097468	10.2598
9	1.83846	.543934	.153486	6.51523	.083486	11.9780
10	1.96715	.508349	.142378	7.02358	.072378	13.8164
11	2.10485	.475093	.133357	7.49867	.063357	15.7836
12	2.25219	.444012	.125902	7.94269	.055902	17.8885
13	2.40985	.414964	.119651	8.35765	.049651	20.1406
14	2.57853	.387817	.114345	8.74547	.044345	22.5505
15	2.75903	.362446	.109795	9.10791	.039795	25.1290
16	2.95216	.338735	.105858	9.44665	.035858	27.8881
17	3.15882	.316574	.102425	9.76322	.032425	30.8402
18	3.37993	.295864	.099413	10.0591	.029413	33.9990
19	3.61653	.276508	.096753	10.3356	.026753	37.3790
20	3.86968	.258419	.094393	10.5940	.024393	40.9955
21	4.14056	.241513	.092289	10.8355	.022289	44.8652
22	4.43040	.225713	.090406	11.0612	.020406	49.0057
23	4.74053	.210947	.088714	11.2722	.018714	53.4361
24	5.07237	.197147	.087189	11.4693	.017189	58.1767
25	5.42743	.184249	.085811	11.6536	.015811	63.2490
26	5.80735	.172195	.084561	11.8258	.014561	68.6765
27	6.21387	.160930	.083426	11.9867	.013426	74.4838
28	6.64884	.150402	.082392	12.1371	.012392	80.6977
29	7.11426	.140563	.081449	12.2777	.011449	87.3465
30	7.61226	.131367	.080586	12.4090	.010586	94.4608
31	8.14511	.122773	.079797	12.5318	.009797	102.073
32	8.71527	.114741	.079073	12.6466	.009073	110.218
33	9.32534	.107235	.078408	12.7538	.008408	118.933
34	9.97811	.100219	.077797	12.8540	.007797	128.259
35	10.6766	.093663	.077234	12.9477	.007234	138.237

8% COMPOUND INTEREST FACTORS

	Single Payment		Uniform Series			
	Compound Amount Factor	Present Worth Factor	Capital Recovery Factor	Present Worth Factor	Sinking Fund Factor	Compound Amount Factor
Periods n	F/P	P/F	A/P	P/A	A/F	F/A
1	1.08000	.925926	1.08000	.925926	1.00000	1.00000
2	1.16640	.857339	.560769	1.78326	.480769	2.08000
3	1.25971	.793832	.388034	2.57710	.308034	3.24640
4	1.36049	.735030	.301921	3.31213	.221921	4.50611
5	1.46933	.680583	.250456	3.99271	.170456	5.86660
6	1.58687	.630170	.216315	4.62288	.136315	7.33593
7	1.71382	.583490	.192072	5.20637	.112072	8.92280
8	1.85093	.540269	.174015	5.74664	.094015	10.6366
9	1.99900	.500249	.160080	6.24689	.080080	12.4876
10	2.15892	.463193	.149029	6.71008	.069029	14.4866
11	2.33164	.428883	.140076	7.13896	.060076	16.6455
12	2.51817	.397114	.132695	7.53608	.052695	18.9771
13	2.71962	.367698	.126522	7.90378	.046522	21.4953
14	2.93719	.340461	.121297	8.24424	.041297	24.2149
15	3.17217	.315242	.116830	8.55948	.036830	27.1521
16	3.42594	.291890	.112977	8.85137	.032977	30.3243
17	3.70002	.270269	.109629	9.12164	.029629	33.7502
18	3.99602	.250249	.106702	9.37189	.026702	37.4502
19	4.31570	.231712	.104128	9.60360	.024128	41.4463
20	4.66096	.214548	.101852	9.81815	.021852	45.7620
21	5.03383	.198656	.099832	10.0168	.019832	50.4229
22	5.43654	.183941	.098032	10.2007	.018032	55.4568
23	5.87146	.170315	.096422	10.3711	.016422	60.8933
24	6.34118	.157699	.094978	10.5288	.014978	66.7648
25	6.84848	.146018	.093679	10.6748	.013679	73.1059
26	7.39635	.135202	.092507	10.8100	.012507	79.9544
27	7.98806	.125187	.091448	10.9352	.011448	87.3508
28	8.62711	.115914	.090489	11.0511	.010489	95.3388
29	9.31727	.107328	.089619	11.1584	.009619	103.966
30	10.0627	.099377	.088827	11.2578	.008827	113.283
31	10.8677	.092016	.088107	11.3498	.008107	123.346
32	11.7371	.085200	.087451	11.4350	.007451	134.214
33	12.6760	.078889	.086852	11.5139	.006852	145.951
34	13.6901	.073045	.086304	11.5869	.006304	158.627
35	14.7853	.067635	.085803	11.6546	.005803	172.317

9% COMPOUND INTEREST FACTORS

	Single Payment		Uniform Series			
	Compound Amount Factor	Present Worth Factor	Capital Recovery Factor	Present Worth Factor	Sinking Fund Factor	Compound Amount Factor
Periods n	F/P	P/F	A/P	P/A	A/F	F/A
1	1.09000	.917431	1.09000	.917431	1.00000	1.00000
2	1.18810	.841680	.568469	1.75911	.478469	2.09000
3	1.29503	.772183	.395055	2.53129	.305055	3.27810
4	1.41158	.708425	.308669	3.23972	.218669	4.57313
5	1.53862	.649931	.257092	3.88965	.167092	5.98471
6	1.67710	.596267	.222920	4.48592	.132920	7.52333
7	1.82804	.547034	.198691	5.03295	.108691	9.20043
8	1.99256	.501866	.180674	5.53482	.090674	11.0285
9	2.17189	.460428	.166799	5.99525	.076799	13.0210
10	2.36736	.422411	.155820	6.41766	.065820	15.1929
11	2.58043	.387533	.146947	6.80519	.056947	17.5603
12	2.81266	.355535	.139651	7.16073	.049651	20.1407
13	3.06580	.326179	.133567	7.48690	.043567	22.9534
14	3.34173	.299246	.128433	7.78615	.038433	26.0192
15	3.64248	.274538	.124059	8.06069	.034059	29.3609
16	3.97031	.251870	.120300	8.31256	.030300	33.0034
17	4.32763	.231073	.117046	8.54363	.027046	36.9737
18	4.71712	.211994	.114212	8.75563	.024212	41.3013
19	5.14166	.194490	.111730	8.95011	.021730	46.0185
20	5.60441	.178431	.109546	9.12855	.019546	51.1601
21	6.10881	.163698	.107617	9.29224	.017617	56.7645
22	6.65860	.150182	.105905	9.44243	.015905	62.8733
23	7.25787	.137781	.104382	9.58021	.014382	69.5319
24	7.91108	.126405	.103023	9.70661	.013023	76.7898
25	8.62308	.115968	.101806	9.82258	.011806	84.7009
26	9.39916	.106393	.100715	9.92897	.010715	93.3240
27	10.2451	.097608	.099735	10.0266	.009735	102.723
28	11.1671	.089548	.098852	10.1161	.008852	112.968
29	12.1722	.082155	.098056	10.1983	.008056	124.135
30	13.2677	.075371	.097336	10.2737	.007336	136.308
31	14.4618	.069148	.096686	10.3428	.006686	149.575
32	15.7633	.063438	.096096	10.4062	.006096	164.037
33	17.1820	.058200	.095562	10.4644	.005562	179.800
34	18.7284	.053395	.095077	10.5178	.005077	196.982
35	20.4140	.048986	.094636	10.5668	.004636	215.711

10% COMPOUND INTEREST FACTORS

	Single Payment		Uniform Series			
	Compound Amount Factor	Present Worth Factor	Capital Recovery Factor	Present Worth Factor	Sinking Fund Factor	Compound Amount Factor
Periods n	F/P	P/F	A/P	P/A	A/F	F/A
1	1.10000	.909091	1.10000	.909091	1.00000	1.00000
2	1.21000	.826446	.576190	1.73554	.476190	2.10000
3	1.33100	.751315	.402115	2.48685	.302115	3.31000
4	1.46410	.683013	.315471	3.16987	.215471	4.64100
5	1.61051	.620921	.263797	3.79079	.163797	6.10510
6	1.77156	.564474	.229607	4.35526	.129607	7.71561
7	1.94872	.513158	.205405	4.86842	.105405	9.48717
8	2.14359	.466507	.187444	5.33493	.087444	11.4359
9	2.35795	.424098	.173641	5.75902	.073641	13.5795
10	2.59374	.385543	.162745	6.14457	.062745	15.9374
11	2.85312	.350494	.153963	6.49506	.053963	18.5312
12	3.13843	.318631	.146763	6.81369	.046763	21.3843
13	3.45227	.289664	.140779	7.10336	.040779	24.5227
14	3.79750	.263331	.135746	7.36669	.035746	27.9750
15	4.17725	.239392	.131474	7.60608	.031474	31.7725
16	4.59497	.217629	.127817	7.82371	.027817	35.9497
17	5.05447	.197845	.124664	8.02155	.024664	40.5447
18	5.55992	.179859	.121930	8.20141	.021930	45.5992
19	6.11591	.163508	.119547	8.36492	.019547	51.1591
20	6.72750	.148644	.117460	8.51356	.017460	57.2750
21	7.40025	.135131	.115624	8.64869	.015624	64.0025
22	8.14027	.122846	.114005	8.77154	.014005	71.4027
23	8.95430	.111678	.112572	8.88322	.012572	79.5430
24	9.84973	.101526	.111300	8.98474	.011300	88.4973
25	10.8347	.092296	.110168	9.07704	.010168	98.3471
26	11.9182	.083905	.109159	9.16095	.009159	109.182
27	13.1100	.076278	.108258	9.23722	.008258	121.100
28	14.4210	.069343	.107451	9.30657	.007451	134.210
29	15.8631	.063039	.106728	9.36961	.006728	148.631
30	17.4494	.057309	.106079	9.42691	.006079	164.494
31	19.1943	.052099	.105496	9.47901	.005496	181.943
32	21.1138	.047362	.104972	9.52638	.004972	201.138
33	23.2252	.043057	.104499	9.56943	.004499	222.252
34	25.5477	.039143	.104074	9.60857	.004074	245.477
35	28.1024	.035584	.103690	9.64416	.003690	271.024

11% COMPOUND INTEREST FACTORS

	Single Payment		Uniform Series			
	Compound Amount Factor	Present Worth Factor	Capital Recovery Factor	Present Worth Factor	Sinking Fund Factor	Compound Amount Factor
Periods n	F/P	P/F	A/P	P/A	A/F	F/A
1	1.11000	.900901	1.11000	.900901	1.00000	1.00000
2	1.23210	.811622	.583934	1.71252	.473934	2.11000
3	1.36763	.731191	.409213	2.44371	.299213	3.34210
4	1.51807	.658731	.322326	3.10245	.212326	4.70973
5	1.68506	.593451	.270570	3.69590	.160570	6.22780
6	1.87041	.534641	.236377	4.23054	.126377	7.91286
7	2.07616	.481658	.212215	4.71220	.102215	9.78327
8	2.30454	.433926	.194321	5.14612	.084321	11.8594
9	2.55804	.390925	.180602	5.53705	.070602	14.1640
10	2.83942	.352184	.169801	5.88923	.059801	16.7220
11	3.15176	.317283	.161121	6.20652	.051121	19.5614
12	3.49845	.285841	.154027	6.49236	.044027	22.7132
13	3.88328	.257514	.148151	6.74987	.038151	26.2116
14	4.31044	.231995	.143228	6.98187	.033228	30.0949
15	4.78459	.209004	.139065	7.19087	.029065	34.4054
16	5.31089	.188292	.135517	7.37916	.025517	39.1899
17	5.89509	.169633	.132471	7.54879	.022471	44.5008
18	6.54355	.152822	.129843	7.70162	.019843	50.3959
19	7.26334	.137678	.127563	7.83929	.017563	56.9395
20	8.06231	.124034	.125576	7.96333	.015576	64.2028
21	8.94917	.111742	.123838	8.07507	.013838	72.2651
22	9.93357	.100669	.122313	8.17574	.012313	81.2143
23	11.0263	.090693	.120971	8.26643	.010971	91.1479
24	12.2392	.081705	.119787	8.34814	.009787	102.174
25	13.5855	.073608	.118740	8.42174	.008740	114.413
26	15.0799	.066314	.117813	8.48806	.007813	127.999
27	16.7386	.059742	.116989	8.54780	.006989	143.079
28	18.5799	.053822	.116257	8.60162	.006257	159.817
29	20.6237	.048488	.115605	8.65011	.005605	178.397
30	22.8923	.043683	.115025	8.69379	.005025	199.021
31	25.4104	.039354	.114506	8.73315	.004506	221.913
32	28.2056	.035454	.114043	8.76860	.004043	247.324
33	31.3082	.031940	.113629	8.80054	.003629	275.529
34	34.7521	.028775	.113259	8.82932	.003259	306.837
35	38.5749	.025924	.112927	8.85524	.002927	341.590

12% COMPOUND INTEREST FACTORS

	Single Payment		Uniform Series			
	Compound Amount Factor	Present Worth Factor	Capital Recovery Factor	Present Worth Factor	Sinking Fund Factor	Compound Amount Factor
Periods n	F/P	P/F	A/P	P/A	A/F	F/A
1	1.12000	.892857	1.12000	.892857	1.00000	1.00000
2	1.25440	.797194	.591698	1.69005	.471698	2.12000
3	1.40493	.711780	.416349	2.40183	.296349	3.37440
4	1.57352	.635518	.329234	3.03735	.209234	4.77933
5	1.76234	.567427	.277410	3.60478	.157410	6.35285
6	1.97382	.506631	.243226	4.11141	.123226	8.11519
7	2.21068	.452349	.219118	4.56376	.099118	10.0890
8	2.47596	.403883	.201303	4.96764	.081303	12.2997
9	2.77308	.360610	.187679	5.32825	.067679	14.7757
10	3.10585	.321973	.176984	5.65022	.056984	17.5487
11	3.47855	.287476	.168415	5.93770	.048415	20.6546
12	3.89598	.256675	.161437	6.19437	.041437	24.1331
13	4.36349	.229174	.155677	6.42355	.035677	28.0291
14	4.88711	.204620	.150871	6.62817	.030871	32.3926
15	5.47357	.182696	.146824	6.81086	.026824	37.2797
16	6.13039	.163122	.143390	6.97399	.023390	42.7533
17	6.86604	.145644	.140457	7.11963	.020457	48.8837
18	7.68997	.130040	.137937	7.24967	.017937	55.7497
19	8.61276	.116107	.135763	7.36578	.015763	63.4397
20	9.64629	.103667	.133879	7.46944	.013879	72.0524
21	10.8038	.092560	.132240	7.56200	.012240	81.6987
22	12.1003	.082643	.130811	7.64465	.010811	92.5026
23	13.5523	.073788	.129560	7.71843	.009560	104.603
24	15.1786	.065882	.128463	7.78432	.008463	118.155
25	17.0001	.058823	.127500	7.84314	.007500	133.334
26	19.0401	.052521	.126652	7.89566	.006652	150.334
27	21.3249	.046894	.125904	7.94255	.005904	169.374
28	23.8839	.041869	.125244	7.98442	.005244	190.699
29	26.7499	.037383	.124660	8.02181	.004660	214.583
30	29.9599	.033378	.124144	8.05518	.004144	241.333
31	33.5551	.029802	.123686	8.08499	.003686	271.293
32	37.5817	.026609	.123280	8.11159	.003280	304.848
33	42.0915	.023758	.122920	8.13535	.002920	342.429
34	47.1425	.021212	.122601	8.15656	.002601	384.521
35	52.7996	.018940	.122317	8.17550	.002317	431.663

13% COMPOUND INTEREST FACTORS

	Single Payment		Uniform Series			
	Compound Amount Factor	Present Worth Factor	Capital Recovery Factor	Present Worth Factor	Sinking Fund Factor	Compound Amount Factor
Periods n	F/P	P/F	A/P	P/A	A/F	F/A
1	1.13000	.884956	1.13000	.884956	1.00000	1.00000
2	1.27690	.783147	.599484	1.66810	.469484	2.13000
3	1.44290	.693050	.423522	2.36115	.293522	3.40690
4	1.63047	.613319	.336194	2.97447	.206194	4.84980
5	1.84244	.542760	.284315	3.51723	.154315	6.48027
6	2.08195	.480319	.250153	3.99755	.120153	8.32271
7	2.35261	.425061	.226111	4.42261	.096111	10.4047
8	2.65844	.376160	.208387	4.79877	.078387	12.7573
9	3.00404	.332885	.194869	5.13166	.064869	15.4157
10	3.39457	.294588	.184290	5.42624	.054290	18.4197
11	3.83586	.260698	.175841	5.68694	.045841	21.8143
12	4.33452	.230706	.168986	5.91765	.038986	25.6502
13	4.89801	.204165	.163350	6.12181	.033350	29.9847
14	5.53475	.180677	.158667	6.30249	.028667	34.8827
15	6.25427	.159891	.154742	6.46238	.024742	40.4175
16	7.06733	.141496	.151426	6.60388	.021426	46.6717
17	7.98608	.125218	.148608	6.72909	.018608	53.7391
18	9.02427	.110812	.146201	6.83991	.016201	61.7251
19	10.1974	.098064	.144134	6.93797	.014134	70.7494
20	11.5231	.086782	.142354	7.02475	.012354	80.9468
21	13.0211	.076798	.140814	7.10155	.010814	92.4699
22	14.7138	.067963	.139479	7.16951	.009479	105.491
23	16.6266	.060144	.138319	7.22966	.008319	120.205
24	18.7881	.053225	.137308	7.28288	.007308	136.831
25	21.2305	.047102	.136426	7.32998	.006426	155.620
26	23.9905	.041683	.135655	7.37167	.005655	176.850
27	27.1093	.036888	.134979	7.40856	.004979	200.841
28	30.6335	.032644	.134387	7.44120	.004387	227.950
29	34.6158	.028889	.133867	7.47009	.003867	258.583
30	39.1159	.025565	.133411	7.49565	.003411	293.199
31	44.2010	.022624	.133009	7.51828	.003009	332.315
32	49.9471	.020021	.132656	7.53830	.002656	376.516
33	56.4402	.017718	.132345	7.55602	.002345	426.463
34	63.7774	.015680	.132071	7.57170	.002071	482.903
35	72.0685	.013876	.131829	7.58557	.001829	546.681

14% COMPOUND INTEREST FACTORS

	Single Payment		Uniform Series			
	Compound Amount Factor	Present Worth Factor	Capital Recovery Factor	Present Worth Factor	Sinking Fund Factor	Compound Amount Factor
Periods n	F/P	P/F	A/P	P/A	A/F	F/A
1	1.14000	.877193	1.14000	.877193	1.00000	1.00000
2	1.29960	.769468	.607290	1.64666	.467290	2.14000
3	1.48154	.674972	.430731	2.32163	.290731	3.43960
4	1.68896	.592080	.343205	2.91371	.203205	4.92114
5	1.92541	.519369	.291284	3.43308	.151284	6.61010
6	2.19497	.455587	.257157	3.88867	.117157	8.53552
7	2.50227	.399637	.233192	4.28830	.093192	10.7305
8	2.85259	.350559	.215570	4.63886	.075570	13.2328
9	3.25195	.307508	.202168	4.94637	.062168	16.0853
10	3.70722	.269744	.191714	5.21612	.051714	19.3373
11	4.22623	.236617	.183394	5.45273	.043394	23.0445
12	4.81790	.207559	.176669	5.66029	.036669	27.2707
13	5.49241	.182069	.171164	5.84236	.031164	32.0887
14	6.26135	.159710	.166609	6.00207	.026609	37.5811
15	7.13794	.140096	.162809	6.14217	.022809	43.8424
16	8.13725	.122892	.159615	6.26506	.019615	50.9804
17	9.27646	.107800	.156915	6.37286	.016915	59.1176
18	10.5752	.094561	.154621	6.46742	.014621	68.3941
19	12.0557	.082948	.152663	6.55037	.012663	78.9692
20	13.7435	.072762	.150986	6.62313	.010986	91.0249
21	15.6676	.063826	.149545	6.68696	.009545	104.768
22	17.8610	.055988	.148303	6.74294	.008303	120.436
23	20.3616	.049112	.147231	6.79206	.007231	138.297
24	23.2122	.043081	.146303	6.83514	.006303	158.659
25	26.4619	.037790	.145498	6.87293	.005498	181.871
26	30.1666	.033149	.144800	6.90608	.004800	208.333
27	34.3899	.029078	.144193	6.93515	.004193	238.499
28	39.2045	.025507	.143664	6.96066	.003664	272.889
29	44.6931	.022375	.143204	6.98304	.003204	312.094
30	50.9502	.019627	.142803	7.00266	.002803	356.787
31	58.0832	.017217	.142453	7.01988	.002453	407.737
32	66.2148	.015102	.142147	7.03498	.002147	465.820
33	75.4849	.013248	.141880	7.04823	.001880	532.035
34	86.0528	.011621	.141646	7.05985	.001646	607.520
35	98.1002	.010194	.141442	7.07005	.001442	693.573

15% COMPOUND INTEREST FACTORS

	Single Payment		Uniform Series			
	Compound Amount Factor	Present Worth Factor	Capital Recovery Factor	Present Worth Factor	Sinking Fund Factor	Compound Amount Factor
Periods n	F/P	P/F	A/P	P/A	A/F	F/A
1	1.15000	.869565	1.15000	.869565	1.00000	1.00000
2	1.32250	.756144	.615116	1.62571	.465116	2.15000
3	1.52088	.657516	.437977	2.28323	.287977	3.47250
4	1.74901	.571753	.350265	2.85498	.200265	4.99337
5	2.01136	.497177	.298316	3.35216	.148316	6.74238
6	2.31306	.432328	.264237	3.78448	.114237	8.75374
7	2.66002	.375937	.240360	4.16042	.090360	11.0668
8	3.05902	.326902	.222850	4.48732	.072850	13.7268
9	3.51788	.284262	.209574	4.77158	.059574	16.7858
10	4.04556	.247185	.199252	5.01877	.049252	20.3037
11	4.65239	.214943	.191069	5.23371	.041069	24.3493
12	5.35025	.186907	.184481	5.42062	.034481	29.0017
13	6.15279	.162528	.179110	5.58315	.029110	34.3519
14	7.07571	.141329	.174688	5.72448	.024688	40.5047
15	8.13706	.122894	.171017	5.84737	.021017	47.5804
16	9.35762	.106865	.167948	5.95423	.017948	55.7175
17	10.7613	.092926	.165367	6.04716	.015367	65.0751
18	12.3755	.080805	.163186	6.12797	.013186	75.8364
19	14.2318	.070265	.161336	6.19823	.011336	88.2118
20	16.3665	.061100	.159761	6.25933	.009761	102.444
21	18.8215	.053131	.158417	6.31246	.008417	118.810
22	21.6447	.046201	.157266	6.35866	.007266	137.632
23	24.8915	.040174	.156278	6.39884	.006278	159.276
24	28.6252	.034934	.155430	6.43377	.005430	184.168
25	32.9190	.030378	.154699	6.46415	.004699	212.793
26	37.8568	.026415	.154070	6.49056	.004070	245.712
27	43.5353	.022970	.153526	6.51353	.003526	283.569
28	50.0656	.019974	.153057	6.53351	.003057	327.104
29	57.5755	.017369	.152651	6.55088	.002651	377.170
30	66.2118	.015103	.152300	6.56598	.002300	434.745
31	76.1435	.013133	.151996	6.57911	.001996	500.957
32	87.5651	.011420	.151733	6.59053	.001733	577.100
33	100.700	.009931	.151505	6.60046	.001505	664.666
34	115.805	.008635	.151307	6.60910	.001307	765.365
35	133.176	.007509	.151135	6.61661	.001135	881.170

16% COMPOUND INTEREST FACTORS

	Single Payment		Uniform Series			
	Compound Amount Factor	Present Worth Factor	Capital Recovery Factor	Present Worth Factor	Sinking Fund Factor	Compound Amount Factor
Periods n	F/P	P/F	A/P	P/A	A/F	F/A
1	1.16000	.862069	1.16000	.862069	1.00000	1.00000
2	1.34560	.743163	.622963	1.60523	.462963	2.16000
3	1.56090	.640658	.445258	2.24589	.285258	3.50560
4	1.81064	.552291	.357375	2.79818	.197375	5.06650
5	2.10034	.476113	.305409	3.27429	.145409	6.87714
6	2.43640	.410442	.271390	3.68474	.111390	8.97748
7	2.82622	.353830	.247613	4.03857	.087613	11.4139
8	3.27841	.305025	.230224	4.34359	.070224	14.2401
9	3.80296	.262953	.217082	4.60654	.057082	17.5185
10	4.41144	.226684	.206901	4.83323	.046901	21.3215
11	5.11726	.195417	.198861	5.02864	.038861	25.7329
12	5.93603	.168463	.192415	5.19711	.032415	30.8502
13	6.88579	.145227	.187184	5.34233	.027184	36.7862
14	7.98752	.125195	.182898	5.46753	.022898	43.6720
15	9.26552	.107927	.179358	5.57546	.019358	51.6595
16	10.7480	.093041	.176414	5.66850	.016414	60.9250
17	12.4677	.080207	.173952	5.74870	.013952	71.6730
18	14.4625	.069144	.171885	5.81785	.011885	84.1407
19	16.7765	.059607	.170142	5.87746	.010142	98.6032
20	19.4608	.051385	.168667	5.92884	.008667	115.380
21	22.5745	.044298	.167416	5.97314	.007416	134.841
22	26.1864	.038188	.166353	6.01133	.006353	157.415
23	30.3762	.032920	.165447	6.04425	.005447	183.601
24	35.2364	.028380	.164673	6.07263	.004673	213.978
25	40.8742	.024465	.164013	6.09709	.004013	249.214
26	47.4141	.021091	.163447	6.11818	.003447	290.088
27	55.0004	.018182	.162963	6.13636	.002963	337.502
28	63.8004	.015674	.162548	6.15204	.002548	392.503
29	74.0085	.013512	.162192	6.16555	.002192	456.303
30	85.8499	.011648	.161886	6.17720	.001886	530.312
31	99.5859	.010042	.161623	6.18724	.001623	616.162
32	115.520	.008657	.161397	6.19590	.001397	715.747
33	134.003	.007463	.161203	6.20336	.001203	831.267
34	155.443	.006433	.161036	6.20979	.001036	965.270
35	180.314	.005546	.160892	6.21534	.000892	1120.71

17% COMPOUND INTEREST FACTORS

	Single Payment		Uniform Series			
	Compound Amount Factor	Present Worth Factor	Capital Recovery Factor	Present Worth Factor	Sinking Fund Factor	Compound Amount Factor
Periods						
n	F/P	P/F	A/P	P/A	A/F	F/A
1	1.17000	.854701	1.17000	.854701	1.00000	1.00000
2	1.36890	.730514	.630829	1.58521	.460829	2.17000
3	1.60161	.624371	.452574	2.20958	.282574	3.53890
4	1.87389	.533650	.364533	2.74324	.194533	5.14051
5	2.19245	.456111	.312564	3.19935	.142564	7.01440
6	2.56516	.389839	.278615	3.58918	.108615	9.20685
7	3.00124	.333195	.254947	3.92238	.084947	11.7720
8	3.51145	.284782	.237690	4.20716	.067690	14.7733
9	4.10840	.243404	.224691	4.45057	.054691	18.2847
10	4.80683	.208037	.214657	4.65860	.044657	22.3931
11	5.62399	.177810	.206765	4.83641	.036765	27.1999
12	6.58007	.151974	.200466	4.98839	.030466	32.8239
13	7.69868	.129892	.195378	5.11828	.025378	39.4040
14	9.00745	.111019	.191230	5.22930	.021230	47.1027
15	10.5387	.094888	.187822	5.32419	.017822	56.1101
16	12.3303	.081101	.185004	5.40529	.015004	66.6488
17	14.4265	.069317	.182662	5.47461	.012662	78.9792
18	16.8790	.059245	.180706	5.53385	.010706	93.4056
19	19.7484	.050637	.179067	5.58449	.009067	110.285
20	23.1056	.043280	.177690	5.62777	.007690	130.033
21	27.0336	.036991	.176530	5.66476	.006530	153.139
22	31.6293	.031616	.175550	5.69637	.005550	180.172
23	37.0062	.027022	.174721	5.72340	.004721	211.801
24	43.2973	.023096	.174019	5.74649	.004019	248.808
25	50.6578	.019740	.173423	5.76623	.003423	292.105
26	59.2697	.016872	.172917	5.78311	.002917	342.763
27	69.3455	.014421	.172487	5.79753	.002487	402.032
28	81.1342	.012325	.172121	5.80985	.002121	471.378
29	94.9271	.010534	.171810	5.82039	.001810	552.512
30	111.065	.009004	.171545	5.82939	.001545	647.439
31	129.946	.007696	.171318	5.83709	.001318	758.504
32	152.036	.006577	.171126	5.84366	.001126	888.449
33	177.883	.005622	.170961	5.84928	.000961	1040.49
34	208.123	.004805	.170821	5.85409	.000821	1218.37
35	243.503	.004107	.170701	5.85820	.000701	1426.49

18% COMPOUND INTEREST FACTORS

	Single Payment		Uniform Series			
	Compound Amount Factor	Present Worth Factor	Capital Recovery Factor	Present Worth Factor	Sinking Fund Factor	Compound Amount Factor
Periods n	F/P	P/F	A/P	P/A	A/F	F/A
1	1.18000	.847458	1.18000	.847458	1.00000	1.00000
2	1.39240	.718184	.638716	1.56564	.458716	2.18000
3	1.64303	.608631	.459924	2.17427	.279924	3.57240
4	1.93878	.515789	.371739	2.69006	.191739	5.21543
5	2.28776	.437109	.319778	3.12717	.139778	7.15421
6	2.69955	.370432	.285910	3.49760	.105910	9.44197
7	3.18547	.313925	.262362	3.81153	.082362	12.1415
8	3.75886	.266038	.245244	4.07757	.065244	15.3270
9	4.43545	.225456	.232395	4.30302	.052395	19.0859
10	5.23384	.191064	.222515	4.49409	.042515	23.5213
11	6.17593	.161919	.214776	4.65601	.034776	28.7551
12	7.28759	.137220	.208628	4.79322	.028628	34.9311
13	8.59936	.116288	.203686	4.90951	.023686	42.2187
14	10.1472	.098549	.199678	5.00806	.019678	50.8180
15	11.9737	.083516	.196403	5.09158	.016403	60.9653
16	14.1290	.070776	.193710	5.16235	.013710	72.9390
17	16.6722	.059980	.191485	5.22233	.011485	87.0680
18	19.6733	.050830	.189639	5.27316	.009639	103.740
19	23.2144	.043077	.188103	5.31624	.008103	123.414
20	27.3930	.036506	.186820	5.35275	.006820	146.628
21	32.3238	.030937	.185746	5.38368	.005746	174.021
22	38.1421	.026218	.184846	5.40990	.004846	206.345
23	45.0076	.022218	.184090	5.43212	.004090	244.487
24	53.1090	.018829	.183454	5.45095	.003454	289.494
25	62.6686	.015957	.182919	5.46691	.002919	342.603
26	73.9490	.013523	.182467	5.48043	.002467	405.272
27	87.2598	.011460	.182087	5.49189	.002087	479.221
28	102.967	.009712	.181765	5.50160	.001765	566.481
29	121.501	.008230	.181494	5.50983	.001494	669.447
30	143.371	.006975	.181264	5.51681	.001264	790.948
31	169.177	.005911	.181070	5.52272	.001070	934.319
32	199.629	.005009	.180906	5.52773	.000906	1103.50
33	235.563	.004245	.180767	5.53197	.000767	1303.13
34	277.964	.003598	.180650	5.53557	.000650	1538.69
35	327.997	.003049	.180550	5.53862	.000550	1816.65

19% COMPOUND INTEREST FACTORS

	Single Payment		Uniform Series			
	Compound Amount Factor	Present Worth Factor	Capital Recovery Factor	Present Worth Factor	Sinking Fund Factor	Compound Amount Factor
Periods n	F/P	P/F	A/P	P/A	A/F	F/A
1	1.19000	.840336	1.19000	.840336	1.00000	1.00000
2	1.41610	.706165	.646621	1.54650	.456621	2.19000
3	1.68516	.593416	.467308	2.13992	.277308	3.60610
4	2.00534	.498669	.378991	2.63859	.188991	5.29126
5	2.38635	.419049	.327050	3.05763	.137050	7.29660
6	2.83976	.352142	.293274	3.40978	.103274	9.68295
7	3.37932	.295918	.269855	3.70570	.079855	12.5227
8	4.02139	.248671	.252885	3.95437	.062885	15.9020
9	4.78545	.208967	.240192	4.16333	.050192	19.9234
10	5.69468	.175602	.230471	4.33893	.040471	24.7089
11	6.77667	.147565	.222891	4.48650	.032891	30.4035
12	8.06424	.124004	.216896	4.61050	.026896	37.1802
13	9.59645	.104205	.212102	4.71471	.022102	45.2445
14	11.4198	.087567	.208235	4.80228	.018235	54.8409
15	13.5895	.073586	.205092	4.87586	.015092	66.2607
16	16.1715	.061837	.202523	4.93770	.012523	79.8502
17	19.2441	.051964	.200414	4.98966	.010414	96.0218
18	22.9005	.043667	.198676	5.03333	.008676	115.266
19	27.2516	.036695	.197238	5.07003	.007238	138.166
20	32.4294	.030836	.196045	5.10086	.006045	165.418
21	38.5910	.025913	.195054	5.12677	.005054	197.847
22	45.9233	.021775	.194229	5.14855	.004229	236.438
23	54.6487	.018299	.193542	5.16685	.003542	282.362
24	65.0320	.015377	.192967	5.18223	.002967	337.010
25	77.3881	.012922	.192487	5.19515	.002487	402.042
26	92.0918	.010859	.192086	5.20601	.002086	479.431
27	109.589	.009125	.191750	5.21513	.001750	571.522
28	130.411	.007668	.191468	5.22280	.001468	681.112
29	155.189	.006444	.191232	5.22924	.001232	811.523
30	184.675	.005415	.191034	5.23466	.001034	966.712
31	219.764	.004550	.190869	5.23921	.000869	1151.39
32	261.519	.003824	.190729	5.24303	.000729	1371.15
33	311.207	.003213	.190612	5.24625	.000612	1632.67
34	370.337	.002700	.190514	5.24895	.000514	1943.88
35	440.701	.002269	.190432	5.25122	.000432	2314.21

20% COMPOUND INTEREST FACTORS

	Single Payment		Uniform Series			
	Compound Amount Factor	Present Worth Factor	Capital Recovery Factor	Present Worth Factor	Sinking Fund Factor	Compound Amount Factor
Periods n	F/P	P/F	A/P	P/A	A/F	F/A
1	1.20000	.833333	1.20000	.833333	1.00000	1.00000
2	1.44000	.694444	.654545	1.52778	.454545	2.20000
3	1.72800	.578704	.474725	2.10648	.274725	3.64000
4	2.07360	.482253	.386289	2.58873	.186289	5.36800
5	2.48832	.401878	.334380	2.99061	.134380	7.44160
6	2.98598	.334898	.300706	3.32551	.100706	9.92992
7	3.58318	.279082	.277424	3.60459	.077424	12.9159
8	4.29982	.232568	.260609	3.83716	.060609	16.4991
9	5.15978	.193807	.248079	4.03097	.048079	20.7989
10	6.19174	.161506	.238523	4.19247	.038523	25.9587
11	7.43008	.134588	.231104	4.32706	.031104	32.1504
12	8.91610	.112157	.225265	4.43922	.025265	39.5805
13	10.6993	.093464	.220620	4.53268	.020620	48.4966
14	12.8392	.077887	.216893	4.61057	.016893	59.1959
15	15.4070	.064905	.213882	4.67547	.013882	72.0351
16	18.4884	.054088	.211436	4.72956	.011436	87.4421
17	22.1861	.045073	.209440	4.77463	.009440	105.931
18	26.6233	.037561	.207805	4.81219	.007805	128.117
19	31.9480	.031301	.206462	4.84350	.006462	154.740
20	38.3376	.026084	.205357	4.86958	.005357	186.688
21	46.0051	.021737	.204444	4.89132	.004444	225.026
22	55.2061	.018114	.203690	4.90943	.003690	271.031
23	66.2474	.015095	.203065	4.92453	.003065	326.237
24	79.4968	.012579	.202548	4.93710	.002548	392.484
25	95.3962	.010483	.202119	4.94759	.002119	471.981
26	114.475	.008735	.201762	4.95632	.001762	567.377
27	137.371	.007280	.201467	4.96360	.001467	681.853
28	164.845	.006066	.201221	4.96967	.001221	819.223
29	197.814	.005055	.201016	4.97472	.001016	984.068
30	237.376	.004213	.200846	4.97894	.000846	1181.88
31	284.852	.003511	.200705	4.98245	.000705	1419.26
32	341.822	.002926	.200587	4.98537	.000587	1704.11
33	410.186	.002438	.200489	4.98781	.000489	2045.93
34	492.224	.002032	.200407	4.98984	.000407	2456.12
35	590.668	.001693	.200339	4.99154	.000339	2948.34

21% COMPOUND INTEREST FACTORS

	Single Payment		Uniform Series			
	Compound Amount Factor	Present Worth Factor	Capital Recovery Factor	Present Worth Factor	Sinking Fund Factor	Compound Amount Factor
Periods n	F/P	P/F	A/P	P/A	A/F	F/A
1	1.21000	.826446	1.21000	.826446	1.00000	1.00000
2	1.46410	.683013	.662489	1.50946	.452489	2.21000
3	1.77156	.564474	.482175	2.07393	.272175	3.67410
4	2.14359	.466507	.393632	2.54044	.183632	5.44566
5	2.59374	.385543	.341765	2.92598	.131765	7.58925
6	3.13843	.318631	.308203	3.24462	.098203	10.1830
7	3.79750	.263331	.285067	3.50795	.075067	13.3214
8	4.59497	.217629	.268415	3.72558	.058415	17.1189
9	5.55992	.179859	.256053	3.90543	.046053	21.7139
10	6.72750	.148644	.246665	4.05408	.036665	27.2738
11	8.14027	.122846	.239411	4.17692	.029411	34.0013
12	9.84973	.101526	.233730	4.27845	.023730	42.1416
13	11.9182	.083905	.229234	4.36235	.019234	51.9913
14	14.4210	.069343	.225647	4.43170	.015647	63.9095
15	17.4494	.057309	.222766	4.48901	.012766	78.3305
16	21.1138	.047362	.220441	4.53637	.010441	95.7799
17	25.5477	.039143	.218555	4.57551	.008555	116.894
18	30.9127	.032349	.217020	4.60786	.007020	142.441
19	37.4043	.026735	.215769	4.63460	.005769	173.354
20	45.2593	.022095	.214745	4.65669	.004745	210.758
21	54.7637	.018260	.213906	4.67495	.003906	256.018
22	66.2641	.015091	.213218	4.69004	.003218	310.781
23	80.1795	.012472	.212652	4.70251	.002652	377.045
24	97.0172	.010307	.212187	4.71282	.002187	457.225
25	117.391	.008519	.211804	4.72134	.001804	554.242
26	142.043	.007040	.211489	4.72838	.001489	671.633
27	171.872	.005818	.211229	4.73420	.001229	813.676
28	207.965	.004809	.211015	4.73901	.001015	985.548
29	251.638	.003974	.210838	4.74298	.000838	1193.51
30	304.482	.003284	.210692	4.74627	.000692	1445.15
31	368.423	.002714	.210572	4.74898	.000572	1749.63
32	445.792	.002243	.210472	4.75122	.000472	2118.06
33	539.408	.001854	.210390	4.75308	.000390	2563.85
34	652.683	.001532	.210322	4.75461	.000322	3103.25
35	789.747	.001266	.210266	4.75588	.000266	3755.94

22% COMPOUND INTEREST FACTORS

	Single Payment		Uniform Series			
	Compound Amount Factor	Present Worth Factor	Capital Recovery Factor	Present Worth Factor	Sinking Fund Factor	Compound Amount Factor
Periods n	F/P	P/F	A/P	P/A	A/F	F/A
1	1.22000	.819672	1.22000	.819672	1.00000	1.00000
2	1.48840	.671862	.670450	1.49153	.450450	2.22000
3	1.81585	.550707	.489658	2.04224	.269658	3.70840
4	2.21533	.451399	.401020	2.49364	.181020	5.52425
5	2.70271	.369999	.349206	2.86364	.129206	7.73958
6	3.29730	.303278	.315764	3.16692	.095764	10.4423
7	4.02271	.248589	.292782	3.41551	.072782	13.7396
8	4.90771	.203761	.276299	3.61927	.056299	17.7623
9	5.98740	.167017	.264111	3.78628	.044111	22.6700
10	7.30463	.136899	.254895	3.92318	.034895	28.6574
11	8.91165	.112213	.247807	4.03540	.027807	35.9620
12	10.8722	.091978	.242285	4.12737	.022285	44.8737
13	13.2641	.075391	.237939	4.20277	.017939	55.7459
14	16.1822	.061796	.234491	4.26456	.014491	69.0100
15	19.7423	.050653	.231738	4.31522	.011738	85.1922
16	24.0856	.041519	.229530	4.35673	.009530	104.935
17	29.3844	.034032	.227751	4.39077	.007751	129.020
18	35.8490	.027895	.226313	4.41866	.006313	158.405
19	43.7358	.022865	.225148	4.44152	.005148	194.254
20	53.3576	.018741	.224202	4.46027	.004202	237.989
21	65.0963	.015362	.223432	4.47563	.003432	291.347
22	79.4175	.012592	.222805	4.48822	.002805	356.443
23	96.8894	.010321	.222294	4.49854	.002294	435.861
24	118.205	.008460	.221877	4.50700	.001877	532.750
25	144.210	.006934	.221536	4.51393	.001536	650.955
26	175.936	.005684	.221258	4.51962	.001258	795.165
27	214.642	.004659	.221030	4.52428	.001030	971.102
28	261.864	.003819	.220843	4.52810	.000843	1185.74
29	319.474	.003130	.220691	4.53123	.000691	1447.61
30	389.758	.002566	.220566	4.53379	.000566	1767.08
31	475.505	.002103	.220464	4.53590	.000464	2156.84
32	580.116	.001724	.220380	4.53762	.000380	2632.34
33	707.741	.001413	.220311	4.53903	.000311	3212.46
34	863.444	.001158	.220255	4.54019	.000255	3920.20
35	1053.40	.000949	.220209	4.54114	.000209	4783.64

23% COMPOUND INTEREST FACTORS

	Single Payment		Uniform Series			
	Compound Amount Factor	Present Worth Factor	Capital Recovery Factor	Present Worth Factor	Sinking Fund Factor	Compound Amount Factor
Periods n	F/P	P/F	A/P	P/A	A/F	F/A
1	1.23000	.813008	1.23000	.813008	1.00000	1.00000
2	1.51290	.660982	.678430	1.47399	.448430	2.23000
3	1.86087	.537384	.497173	2.01137	.267173	3.74290
4	2.28887	.436897	.408451	2.44827	.178451	5.60377
5	2.81531	.355201	.356700	2.80347	.126700	7.89263
6	3.46283	.288781	.323389	3.09225	.093389	10.7079
7	4.25928	.234782	.300568	3.32704	.070568	14.1708
8	5.23891	.190879	.284259	3.51792	.054259	18.4300
9	6.44386	.155187	.272249	3.67310	.042249	23.6690
10	7.92595	.126168	.263208	3.79927	.033208	30.1128
11	9.74891	.102576	.256289	3.90185	.026289	38.0388
12	11.9912	.083395	.250926	3.98524	.020926	47.7877
13	14.7491	.067801	.246728	4.05304	.016728	59.7788
14	18.1414	.055122	.243418	4.10816	.013418	74.5280
15	22.3140	.044815	.240791	4.15298	.010791	92.6694
16	27.4462	.036435	.238697	4.18941	.008697	114.983
17	33.7588	.029622	.237021	4.21904	.007021	142.430
18	41.5233	.024083	.235676	4.24312	.005676	176.188
19	51.0737	.019580	.234593	4.26270	.004593	217.712
20	62.8206	.015918	.233720	4.27862	.003720	268.785
21	77.2694	.012942	.233016	4.29156	.003016	331.606
22	95.0413	.010522	.232446	4.30208	.002446	408.875
23	116.901	.008554	.231984	4.31063	.001984	503.917
24	143.788	.006955	.231611	4.31759	.001611	620.817
25	176.859	.005654	.231308	4.32324	.001308	764.605
26	217.537	.004597	.231062	4.32784	.001062	941.465
27	267.570	.003737	.230863	4.33158	.000863	1159.00
28	329.112	.003038	.230701	4.33462	.000701	1426.57
29	404.807	.002470	.230570	4.33709	.000570	1755.68
30	497.913	.002008	.230463	4.33909	.000463	2160.49
31	612.433	.001633	.230376	4.34073	.000376	2658.40
32	753.292	.001328	.230306	4.34205	.000306	3270.84
33	926.550	.001079	.230249	4.34313	.000249	4024.13
34	1139.66	.000877	.230202	4.34401	.000202	4950.68
35	1401.78	.000713	.230164	4.34472	.000164	6090.33

24% COMPOUND INTEREST FACTORS

	Single Payment		Uniform Series			
	Compound Amount Factor	Present Worth Factor	Capital Recovery Factor	Present Worth Factor	Sinking Fund Factor	Compound Amount Factor
Periods n	F/P	P/F	A/P	P/A	A/F	F/A
1	1.24000	.806452	1.24000	.806452	1.00000	1.00000
2	1.53760	.650364	.686429	1.45682	.446429	2.24000
3	1.90662	.524487	.504718	1.98130	.264718	3.77760
4	2.36421	.422974	.415926	2.40428	.175926	5.68422
5	2.93163	.341108	.364248	2.74538	.124248	8.04844
6	3.63522	.275087	.331074	3.02047	.091074	10.9801
7	4.50767	.221844	.308422	3.24232	.068422	14.6153
8	5.58951	.178907	.292293	3.42122	.052293	19.1229
9	6.93099	.144280	.280465	3.56550	.040465	24.7125
10	8.59443	.116354	.271602	3.68186	.031602	31.6434
11	10.6571	.093834	.264852	3.77569	.024852	40.2379
12	13.2148	.075673	.259648	3.85136	.019648	50.8950
13	16.3863	.061026	.255598	3.91239	.015598	64.1097
14	20.3191	.049215	.252423	3.96160	.012423	80.4961
15	25.1956	.039689	.249919	4.00129	.009919	100.815
16	31.2426	.032008	.247936	4.03330	.007936	126.011
17	38.7408	.025813	.246359	4.05911	.006359	157.253
18	48.0386	.020817	.245102	4.07993	.005102	195.994
19	59.5679	.016788	.244098	4.09672	.004098	244.033
20	73.8641	.013538	.243294	4.11026	.003294	303.601
21	91.5915	.010918	.242649	4.12117	.002649	377.465
22	113.574	.008805	.242132	4.12998	.002132	469.056
23	140.831	.007101	.241716	4.13708	.001716	582.630
24	174.631	.005726	.241382	4.14281	.001382	723.461
25	216.542	.004618	.241113	4.14742	.001113	898.092
26	268.512	.003724	.240897	4.15115	.000897	1114.63
27	332.955	.003003	.240723	4.15415	.000723	1383.15
28	412.864	.002422	.240583	4.15657	.000583	1716.10
29	511.952	.001953	.240470	4.15853	.000470	2128.96
30	634.820	.001575	.240379	4.16010	.000379	2640.92
31	787.177	.001270	.240305	4.16137	.000305	3275.74
32	976.099	.001024	.240246	4.16240	.000246	4062.91
33	1210.36	.000826	.240198	4.16322	.000198	5039.01
34	1500.85	.000666	.240160	4.16389	.000160	6249.38
35	1861.05	.000537	.240129	4.16443	.000129	7750.23

25% COMPOUND INTEREST FACTORS

	Single Payment		Uniform Series			
	Compound Amount Factor	Present Worth Factor	Capital Recovery Factor	Present Worth Factor	Sinking Fund Factor	Compound Amount Factor
Periods n	F/P	P/F	A/P	P/A	A/F	F/A
1	1.25000	.800000	1.25000	.800000	1.00000	1.00000
2	1.56250	.640000	.694444	1.44000	.444444	2.25000
3	1.95313	.512000	.512295	1.95200	.262295	3.81250
4	2.44141	.409600	.423442	2.36160	.173442	5.76562
5	3.05176	.327680	.371847	2.68928	.121847	8.20703
6	3.81470	.262144	.338819	2.95142	.088819	11.2588
7	4.76837	.209715	.316342	3.16114	.066342	15.0735
8	5.96046	.167772	.300399	3.32891	.050399	19.8419
9	7.45058	.134218	.288756	3.46313	.038756	25.8023
10	9.31323	.107374	.280073	3.57050	.030073	33.2529
11	11.6415	.085899	.273493	3.65640	.023493	42.5661
12	14.5519	.068719	.268448	3.72512	.018448	54.2077
13	18.1899	.054976	.264543	3.78010	.014543	68.7596
14	22.7374	.043980	.261501	3.82408	.011501	86.9495
15	28.4217	.035184	.259117	3.85926	.009117	109.687
16	35.5271	.028147	.257241	3.88741	.007241	138.109
17	44.4089	.022518	.255759	3.90993	.005759	173.636
18	55.5112	.018014	.254586	3.92794	.004586	218.045
19	69.3889	.014412	.253656	3.94235	.003656	273.556
20	86.7362	.011529	.252916	3.95388	.002916	342.945
21	108.420	.009223	.252327	3.96311	.002327	429.681
22	135.525	.007379	.251858	3.97049	.001858	538.101
23	169.407	.005903	.251485	3.97639	.001485	673.626
24	211.758	.004722	.251186	3.98111	.001186	843.033
25	264.698	.003778	.250948	3.98489	.000948	1054.79
26	330.872	.003022	.250758	3.98791	.000758	1319.49
27	413.590	.002418	.250606	3.99033	.000606	1650.36
28	516.988	.001934	.250485	3.99226	.000485	2063.95
29	646.235	.001547	.250387	3.99381	.000387	2580.94
30	807.794	.001238	.250310	3.99505	.000310	3227.17
31	1009.74	.000990	.250248	3.99604	.000248	4034.97
32	1262.18	.000792	.250198	3.99683	.000198	5044.71
33	1577.72	.000634	.250159	3.99746	.000159	6306.89
34	1972.15	.000507	.250127	3.99797	.000127	7884.61
35	2465.19	.000406	.250101	3.99838	.000101	9856.76

26% COMPOUND INTEREST FACTORS

	Single Payment		Uniform Series			
	Compound Amount Factor	Present Worth Factor	Capital Recovery Factor	Present Worth Factor	Sinking Fund Factor	Compound Amount Factor
Periods n	F/P	P/F	A/P	P/A	A/F	F/A
1	1.26000	.793651	1.26000	.793651	1.00000	1.00000
2	1.58760	.629882	.702478	1.42353	.442478	2.26000
3	2.00038	.499906	.519902	1.92344	.259902	3.84760
4	2.52047	.396751	.430999	2.32019	.170999	5.84798
5	3.17580	.314882	.379496	2.63507	.119496	8.36845
6	4.00150	.249906	.346623	2.88498	.086623	11.5442
7	5.04190	.198338	.324326	3.08331	.064326	15.5458
8	6.35279	.157411	.308573	3.24073	.048573	20.5876
9	8.00451	.124930	.297119	3.36566	.037119	26.9404
10	10.0857	.099150	.288616	3.46481	.028616	34.9449
11	12.7080	.078691	.282207	3.54350	.022207	45.0306
12	16.0120	.062453	.277319	3.60595	.017319	57.7386
13	20.1752	.049566	.273559	3.65552	.013559	73.7506
14	25.4207	.039338	.270647	3.69485	.010647	93.9258
15	32.0301	.031221	.268379	3.72607	.008379	119.347
16	40.3579	.024778	.266606	3.75085	.006606	151.377
17	50.8510	.019665	.265216	3.77052	.005216	191.735
18	64.0722	.015607	.264122	3.78613	.004122	242.585
19	80.7310	.012387	.263261	3.79851	.003261	306.658
20	101.721	.009831	.262581	3.80834	.002581	387.389
21	128.169	.007802	.262045	3.81615	.002045	489.110
22	161.492	.006192	.261620	3.82234	.001620	617.278
23	203.480	.004914	.261284	3.82725	.001284	778.771
24	256.385	.003900	.261018	3.83115	.001018	982.251
25	323.045	.003096	.260807	3.83425	.000807	1238.64
26	407.037	.002457	.260640	3.83670	.000640	1561.68
27	512.867	.001950	.260508	3.83865	.000508	1968.72
28	646.212	.001547	.260403	3.84020	.000403	2481.59
29	814.228	.001228	.260320	3.84143	.000320	3127.80
30	1025.93	.000975	.260254	3.84240	.000254	3942.03
31	1292.67	.000774	.260201	3.84318	.000201	4967.95
32	1628.76	.000614	.260160	3.84379	.000160	6260.62
33	2052.24	.000487	.260127	3.84428	.000127	7889.38
34	2585.82	.000387	.260101	3.84467	.000101	9941.62
35	3258.14	.000307	.260080	3.84497	.000080	12527.4

27% COMPOUND INTEREST FACTORS

	Single Payment		Uniform Series			
	Compound Amount Factor	Present Worth Factor	Capital Recovery Factor	Present Worth Factor	Sinking Fund Factor	Compound Amount Factor
Periods n	F/P	P/F	A/P	P/A	A/F	F/A
1	1.27000	.787402	1.27000	.787402	1.00000	1.00000
2	1.61290	.620001	.710529	1.40740	.440529	2.27000
3	2.04838	.488190	.527539	1.89559	.257539	3.88290
4	2.60145	.384402	.438598	2.27999	.168598	5.93128
5	3.30384	.302678	.387196	2.58267	.117196	8.53273
6	4.19587	.238329	.354484	2.82100	.084484	11.8366
7	5.32876	.187661	.332374	3.00866	.062374	16.0324
8	6.76752	.147765	.316814	3.15643	.046814	21.3612
9	8.59475	.116350	.305551	3.27278	.035551	28.1287
10	10.9153	.091614	.297231	3.36439	.027231	36.7235
11	13.8625	.072137	.290991	3.43653	.020991	47.6388
12	17.6053	.056801	.286260	3.49333	.016260	61.5013
13	22.3588	.044725	.282641	3.53806	.012641	79.1066
14	28.3957	.035217	.279856	3.57327	.009856	101.465
15	36.0625	.027730	.277701	3.60100	.007701	129.861
16	45.7994	.021834	.276027	3.62284	.006027	165.924
17	58.1652	.017192	.274723	3.64003	.004723	211.723
18	73.8698	.013537	.273705	3.65357	.003705	269.888
19	93.8147	.010659	.272909	3.66422	.002909	343.758
20	119.145	.008393	.272285	3.67262	.002285	437.573
21	151.314	.006609	.271796	3.67923	.001796	556.717
22	192.168	.005204	.271412	3.68443	.001412	708.031
23	244.054	.004097	.271111	3.68853	.001111	900.199
24	309.948	.003226	.270874	3.69175	.000874	1144.25
25	393.634	.002540	.270688	3.69429	.000688	1454.20
26	499.916	.002000	.270541	3.69630	.000541	1847.84
27	634.893	.001575	.270426	3.69787	.000426	2347.75
28	806.314	.001240	.270335	3.69911	.000335	2982.64
29	1024.02	.000977	.270264	3.70009	.000264	3788.96
30	1300.50	.000769	.270208	3.70086	.000208	4812.98
31	1651.64	.000605	.270164	3.70146	.000164	6113.48
32	2097.58	.000477	.270129	3.70194	.000129	7765.12
33	2663.93	.000375	.270101	3.70231	.000101	9862.70
34	3383.19	.000296	.270080	3.70261	.000080	12526.6
35	4296.65	.000233	.270063	3.70284	.000063	15909.8

28% COMPOUND INTEREST FACTORS

	Single Payment		Uniform Series			
	Compound Amount Factor	Present Worth Factor	Capital Recovery Factor	Present Worth Factor	Sinking Fund Factor	Compound Amount Factor
Periods n	F/P	P/F	A/P	P/A	A/F	F/A
1	1.28000	.781250	1.28000	.781250	1.00000	1.00000
2	1.63840	.610352	.718596	1.39160	.438596	2.28000
3	2.09715	.476837	.535206	1.86844	.255206	3.91840
4	2.68435	.372529	.446236	2.24097	.166236	6.01555
5	3.43597	.291038	.394944	2.53201	.114944	8.69991
6	4.39805	.227374	.362400	2.75938	.082400	12.1359
7	5.62950	.177636	.340482	2.93702	.060482	16.5339
8	7.20576	.138778	.325119	3.07579	.045119	22.1634
9	9.22337	.108420	.314049	3.18421	.034049	29.3692
10	11.8059	.084703	.305912	3.26892	.025912	38.5926
11	15.1116	.066174	.299842	3.33509	.019842	50.3985
12	19.3428	.051699	.295265	3.38679	.015265	65.5100
13	24.7588	.040390	.291785	3.42718	.011785	84.8529
14	31.6913	.031554	.289123	3.45873	.009123	109.612
15	40.5648	.024652	.287077	3.48339	.007077	141.303
16	51.9230	.019259	.285499	3.50265	.005499	181.868
17	66.4614	.015046	.284277	3.51769	.004277	233.791
18	85.0706	.011755	.283331	3.52945	.003331	300.252
19	108.890	.009184	.282595	3.53863	.002595	385.323
20	139.380	.007175	.282023	3.54580	.002023	494.213
21	178.406	.005605	.281578	3.55141	.001578	633.593
22	228.360	.004379	.281232	3.55579	.001232	811.999
23	292.300	.003421	.280961	3.55921	.000961	1040.36
24	374.144	.002673	.280750	3.56188	.000750	1332.66
25	478.905	.002088	.280586	3.56397	.000586	1706.80
26	612.998	.001631	.280458	3.56560	.000458	2185.71
27	784.638	.001274	.280357	3.56688	.000357	2798.71
28	1004.34	.000996	.280279	3.56787	.000279	3583.34
29	1285.55	.000778	.280218	3.56865	.000218	4587.68
30	1645.50	.000608	.280170	3.56926	.000170	5873.23
31	2106.25	.000475	.280133	3.56973	.000133	7518.74
32	2695.99	.000371	.280104	3.57010	.000104	9624.98
33	3450.87	.000290	.280081	3.57039	.000081	12321.0
34	4417.12	.000226	.280063	3.57062	.000063	15771.8
35	5653.91	.000177	.280050	3.57080	.000050	20189.0

29% COMPOUND INTEREST FACTORS

	Single Payment		Uniform Series			
	Compound Amount Factor	Present Worth Factor	Capital Recovery Factor	Present Worth Factor	Sinking Fund Factor	Compound Amount Factor
Periods n	F/P	P/F	A/P	P/A	A/F	F/A
1	1.29000	.775194	1.29000	.775194	1.00000	1.00000
2	1.66410	.600925	.726681	1.37612	.436681	2.29000
3	2.14669	.465834	.542902	1.84195	.252902	3.95410
4	2.76923	.361111	.453913	2.20306	.163913	6.10079
5	3.57231	.279931	.402739	2.48300	.112739	8.87002
6	4.60827	.217001	.370371	2.70000	.080371	12.4423
7	5.94467	.168218	.348649	2.86821	.058649	17.0506
8	7.66863	.130401	.333487	2.99862	.043487	22.9953
9	9.89253	.101086	.322612	3.09970	.032612	30.6639
10	12.7614	.078362	.314657	3.17806	.024657	40.5564
11	16.4622	.060745	.308755	3.23881	.018755	53.3178
12	21.2362	.047089	.304331	3.28590	.014331	69.7800
13	27.3947	.036503	.300987	3.32240	.010987	91.0161
14	35.3391	.028297	.298445	3.35070	.008445	118.411
15	45.5875	.021936	.296504	3.37264	.006504	153.750
16	58.8079	.017005	.295017	3.38964	.005017	199.337
17	75.8621	.013182	.293874	3.40282	.003874	258.145
18	97.8622	.010218	.292994	3.41304	.002994	334.007
19	126.242	.007921	.292316	3.42096	.002316	431.870
20	162.852	.006141	.291792	3.42710	.001792	558.112
21	210.080	.004760	.291387	3.43186	.001387	720.964
22	271.003	.003690	.291074	3.43555	.001074	931.044
23	349.593	.002860	.290832	3.43841	.000832	1202.05
24	450.976	.002217	.290644	3.44063	.000644	1551.64
25	581.759	.001719	.290499	3.44235	.000499	2002.62
26	750.468	.001333	.290387	3.44368	.000387	2584.37
27	968.104	.001033	.290300	3.44471	.000300	3334.84
28	1248.85	.000801	.290232	3.44551	.000232	4302.95
29	1611.02	.000621	.290180	3.44614	.000180	5551.80
30	2078.22	.000481	.290140	3.44662	.000140	7162.82
31	2680.90	.000373	.290108	3.44699	.000108	9241.04
32	3458.36	.000289	.290084	3.44728	.000084	11921.9
33	4461.29	.000224	.290065	3.44750	.000065	15380.3
34	5755.06	.000174	.290050	3.44768	.000050	19841.6
35	7424.03	.000135	.290039	3.44781	.000039	25596.7

30% COMPOUND INTEREST FACTORS

	Single Payment		Uniform Series			
	Compound Amount Factor	Present Worth Factor	Capital Recovery Factor	Present Worth Factor	Sinking Fund Factor	Compound Amount Factor
Periods n	F/P	P/F	A/P	P/A	A/F	F/A
1	1.30000	.769231	1.30000	.769231	1.00000	1.00000
2	1.69000	.591716	.734783	1.36095	.434783	2.30000
3	2.19700	.455166	.550627	1.81611	.250627	3.99000
4	2.85610	.350128	.461629	2.16624	.161629	6.18700
5	3.71293	.269329	.410582	2.43557	.110582	9.04310
6	4.82681	.207176	.378394	2.64275	.078394	12.7560
7	6.27485	.159366	.356874	2.80211	.056874	17.5828
8	8.15731	.122589	.341915	2.92470	.041915	23.8577
9	10.6045	.094300	.331235	3.01900	.031235	32.0150
10	13.7858	.072538	.323463	3.09154	.023463	42.6195
11	17.9216	.055799	.317729	3.14734	.017729	56.4053
12	23.2981	.042922	.313454	3.19026	.013454	74.3270
13	30.2875	.033017	.310243	3.22328	.010243	97.6250
14	39.3738	.025398	.307818	3.24867	.007818	127.913
15	51.1859	.019537	.305978	3.26821	.005978	167.286
16	66.5417	.015028	.304577	3.28324	.004577	218.472
17	86.5042	.011560	.303509	3.29480	.003509	285.014
18	112.455	.008892	.302692	3.30369	.002692	371.518
19	146.192	.006840	.302066	3.31053	.002066	483.973
20	190.050	.005262	.301587	3.31579	.001587	630.165
21	247.065	.004048	.301219	3.31984	.001219	820.215
22	321.184	.003113	.300937	3.32296	.000937	1067.28
23	417.539	.002395	.300720	3.32535	.000720	1388.46
24	542.801	.001842	.300554	3.32719	.000554	1806.00
25	705.641	.001417	.300426	3.32861	.000426	2348.80
26	917.333	.001090	.300327	3.32970	.000327	3054.44
27	1192.53	.000839	.300252	3.33054	.000252	3971.78
28	1550.29	.000645	.300194	3.33118	.000194	5164.31
29	2015.38	.000496	.300149	3.33168	.000149	6714.60
30	2620.00	.000382	.300115	3.33206	.000115	8729.99
31	3405.99	.000294	.300088	3.33235	.000088	11350.0
32	4427.79	.000226	.300068	3.33258	.000068	14756.0
33	5756.13	.000174	.300052	3.33275	.000052	19183.8
34	7482.97	.000134	.300040	3.33289	.000040	24939.9
35	9727.86	.000103	.300031	3.33299	.000031	32422.9

Index

A

Advisory group, 15, 149

Alternate energy sources, 79

Audience make-up, 132

Audio visuals, 136

Audit instruments, 23
    planning, 23
    team, 39
    types, 37

Awards program, 71

B

Bar chart, 25

Basic interest, 107

Blackboards, 137

Budget, 17

C

Capacitive load, 171

Capital recovery factor, 112

Check list, 41

Committee, 15, 194

Conservation behavior, 65

Contingency planning, 75

Continuing education programs, 1, 47

Coordinators, 16, 194
    educational plans, 48

Cost per million BTU's, 35

D

Data collection, 39

Degree days, 10

Demand control, 161, 219

Demand, 158, 179

Discipline of energy management, 1

Disencentives, 65

Disruptions, 75

E

ECO's, analyzing, 42
    identifying, 41

Economic evaluation, 97

Educational plan, 47, 70, 149

Electric utility, 25

Electrical costs, 155

Electrical survey, 187

Employee involvement, 59
    lay-off plan, 80
    motivation ideas, 69
    educational plans, 52

Energy Management Diploma Program, 4

Energy committee, 70

Energy history, 9

Energy index, 10

Energy intensive products 31

Energy management system, 203

Energy manager, 14

Energy relationships, 31

Energy security, 75

Energy services, 101

Energy, definition, 158

Equivalence, 105

F

Financing schemes, 98

Flip charts, 138

Forecasting, 11

Future energy costs, 127

G

Group dynamics, 91

Group member characteristics, 92

H

Heating values, 9

Hurdle rate, 36

I

Incentives, 65

Inductive load, 168

Instruments, 188
    audit, 223

Insurance, 102

Internal rate of return, 122

L

Ladder of priorities, 61

Leasing, 102

Life cycle costing, 104

Line graphs, 25

Load factor, 179

Load shedding, 161

Logo, 49

M

Management, educational plans, 51

Manual, energy, 49

Micro-audits, 72

Motivation, 4

Motor horsepower, 189

N

Networking, 147

New technology, 220

Nominal Group Technique, 8, 70, 94, 131, 149

O

OPEC, 127

Organiation, 13, 200

Organiational planning, 7

Overhead projectors, 138

P

Payback, 36, 121

Peak shaving, 206

Persuasive communications, 64

Pie chart, 25

Planning, 4

Policy, energy, 11
    sample, 193

Positive cash flow financing project, 103

Power factor, 171, 179

Power, electrical, 157

Pre-audit planning, 24

Presentation, 133

Presenting your ideas, 131

Priority list, 23, 43
    sample, 227

Psychology of conservation, 63

Psychology of motivation, 60

R

Rate of return, 122

Rate schedules, 156

Rate structure, 176

Reporting, 194

Reports, 19

Resistive load, 167

Return on investment, (ROI) 36

S

Selecting an energy manager, 14

Selling the program, 11

Seminar topics, 55

Shared savings financing, 99

Shutdown priority, 78

Simulators, 71

Single line utility diagram, 31

Single payment compound factor, 109

Sinking fund factor, 114

Slide presentation, 21

Social incentives, 67

Strategic plan, sample, 233

Strategic planning format, 85

Strategic planning, 77, 83

Supplier evaluation, 80

Synchronous motors, 176

T

Tax on energy, 31

Technical staff, 15

Time value of money, 105

Training, 194

Tuning, operation, and maintenance, (TOM), 37

U

Uniform series worth factor, 113

Utility curtailment plan, 77, 82

Utilization factor, 179

W

Workshop topics, 56, 58

Workshops, 93